MODERN TRIGONOMETRY

BY

GEF ARMSTRONG
ELLA WRIGHT
AVA WHITE

KNOWLEDGE BAKERS

KNOWLEDGE BAKERS

ISBN: 978-81-97559-42-6 (HB)

Published by : **Knowledge Bakers**
Pune, India
Email: info@knowledgebakers.com

PREFACE

Trigonometry, the branch of mathematics concerned with the study of angles and the relationships between them, has been a cornerstone of mathematical inquiry for centuries. From ancient civilizations to modern scientific endeavors, trigonometry has played a crucial role in a wide range of fields, including astronomy, engineering, physics, and surveying. However, in today's rapidly evolving technological landscape, trigonometry is experiencing a resurgence in relevance and applicability, with new computational tools, data analysis techniques, and interdisciplinary connections expanding its reach into diverse areas of inquiry. As we embark on this exploration of modern trigonometry, we delve into a realm where mathematical concepts, computational methods, and real-world applications converge, uncovering the contemporary relevance and expanding horizons of this foundational branch of mathematics.

This book, titled "Modern Trigonometry," is crafted to provide a comprehensive overview of trigonometric principles, techniques, and applications in the context of contemporary mathematics and scientific inquiry. Through collaboration with leading mathematicians, educators, and researchers in the field of trigonometry, we aim to offer readers a multidimensional and insightful journey through the intricacies of modern trigonometry.

As editors of this volume, we have curated a collection of chapters that cover a wide range of topics in modern trigonometry, including trigonometric functions, identities, equations, vectors, complex numbers, Fourier analysis, and applications in fields such as physics, engineering, computer graphics, and signal processing. Each chapter is designed to provide readers with a deep understanding of the theoretical foundations, computational techniques, and practical applications of trigonometry in contemporary contexts.

Furthermore, this book seeks to highlight the interdisciplinary nature of modern trigonometry and its connections to other branches of mathematics and scientific inquiry. By exploring the role of trigonometry in solving real-world problems, modeling physical phenomena, and advancing technological innovations, we aim to inspire readers to appreciate the beauty and utility of trigonometric concepts in the modern age.

As we embark on this enlightening journey through the world of modern trigonometry, we extend our gratitude to the contributors whose expertise and dedication have enriched this book. We hope that "Modern Trigonometry" serves as a valuable resource for students, educators, researchers, and enthusiasts alike, empowering them to deepen their understanding of trigonometric principles and to explore the myriad applications and interdisciplinary connections of trigonometry in contemporary mathematics and scientific inquiry.

Gef Armstrong
Ella Wright
Ava White

Contents

Chapter 1

Right Triangle Trigonometry

The Origins of Trigonometry

The precursors to what we study today as Trigonometry had their origin in ancient Mesopotamia, Greece and India. These cultures used the concepts of angles and lengths as an aid to understanding the movements of the heavenly bodies in the night sky. Ancient trigonometry typically used angles and triangles that were embedded in circles so that many of the calculations used were based on the lengths of chords within a circle. The relationships between the lengths of the chords and other lines drawn within a circle and the measure of the corresponding central angle represent the foundation of trigonometry - the relationship between angles and distances.

The earliest values for the sine function were calculated by Indian mathematicians in the 5th century. The cosine and tangent, as well as the cotangent, secant and cosecant were developed by Islamic mathematicians by the 11th century. European navigators used these ideas extensively to help calculate distances and direction during the Middle Ages. Modern European trigonometry as we understand it was then developed throughout the Renaissance (1450-1650) and Enlightenment (1650-1800).

1.1 Measuring Angles

Measuring Angles in Degrees

The two most common units for measuring angles are degrees and radians. Degrees are based on the ancient Mesopotamian assignment of 360° to a complete circle. This has its origin in the division of the horizon of the nighttime sky as the earth takes 365 days to travel around the sun. Because degrees were originally developed by the Mesopotamians, they are often also broken out into 60 unit measures of minutes and seconds. Sixty seconds make one minute and sixty minutes makes one degree.

60 seconds=1 minute
or
$60''=1'$
60 minutes=1 degree
or
$60'=1^\circ$

Angles measured in degrees may also be expressed using decimal portions of a degree, for example:

$$72.5^\circ=72^\circ\ 30'$$

Converting from decimal to DMS

Converting between degrees expressed with decimals and the degrees, minutes, seconds format (DMS) is relatively simple. If you're converting from degrees expressed with decimals to DMS, simply take the portion of the angle behind the decimal point and multiply by 60. In our previous example, we would take the .5 from 72.5°and multiply this by 60: 0.5*60=30. So, the angle in DMS units would be $72^\circ\ 30'$.

Examples

Convert 21.85°to DMS units

$0.85 * 60 = 51$

So, 21.85°=21° 51′

Convert 143.27°to DMS units

$0.27 * 60 = 16.2$

So, 143.27° = 143° 16.2′

In order to compute the number of seconds needed to express this angle in DMS units, we take the decimal portion of the minutes and multiply by 60:

$0.2 * 60 = 12$

So, 143.27° = 143° 16.2′ = 143° 16′ 12″

In the example above we ended with a whole number of seconds. If you don't get a whole number for the seconds then you can leave the seconds with a decimal portion. For example, if you wanted to convert 22.847°to DMS units:

22.847° = 22° 50.82′ = 22° 50′ 49.2″

Converting from DMS to decimal

To convert from DMS units to decimals, simply take the seconds portion and divide by 60 to make it a decimal:

129° 19′ 30″ = 129° 19.5′

Then take the new minutes portion and divide it by 60

$\frac{19.5}{60} = 0.325$

This is the decimal portion of the angle

$129^\circ\ 19'\ 30'' = 129^\circ\ 19.5' = 129.325^\circ$

If you end up with repeating decimals in this process that's fine - just indicate the repeating portion with a bar.

Examples

Convert $42^\circ\ 27'\ 36''$ to decimal degrees

$\frac{36}{60} = 0.6$

$42^\circ\ 27'\ 36'' = 42^\circ\ 27.6'$

$\frac{27.6}{60} = 0.46$

$42^\circ\ 27.6' = 42.46^\circ$

Convert $17^\circ\ 40'\ 18''$ to decimal degrees

$\frac{18}{60} = 0.3$

$17^\circ\ 40'\ 18'' = 17^\circ\ 40.3'$

$\frac{40.3}{60} = 0.671\overline{6}$

$17^\circ\ 40'\ 18'' = 17^\circ\ 40.3' = 17.671\overline{6}^\circ$

Measuring Angles in Radians

The other most commonly used method for measuring angles is radian measure. Radian measure is based on the central angle of a circle. A given central angle will trace out an arc of a particular length on the circle. The ratio of the arc length to the radius of the circle is the angle measure in radians. The benefit of radian measure is that it is based on a ratio of distances whereas degree measure is not. This allows radians to be used in calculus in situations in which degree measure would be inappropriate.

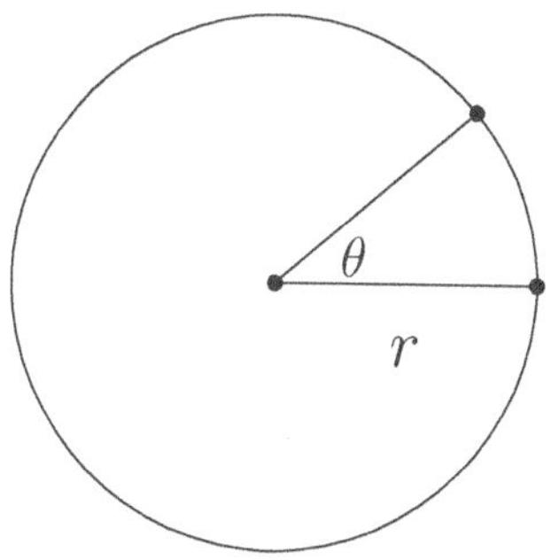

The length of the arc intersected by the central angle is the portion of the circumference swept out by the angle along the edge of the circle. The circumference of the circle would be $2\pi r$, so the length of the arc would be $\frac{\theta}{360^\circ} * 2\pi r$. The ratio of this arclength to the radius is $\frac{\frac{\theta}{360^\circ} * 2\pi r}{r}$ or

$$\frac{2\pi}{360^\circ} * \theta$$

or in reduced form

$$\frac{\pi}{180^\circ} * \theta$$

This assumes that the angle has been expressed in degrees to begin with. If an angle is expressed in radian measure, then to convert it into degrees, simply multiply by $\frac{180^\circ}{\pi}$.

Examples - Degrees to Radians

Convert 60° to radians

$\frac{\pi}{180^\circ} * 60^\circ = \frac{\pi}{3}$

Convert 142° to radians

$\frac{\pi}{180^\circ} * 142^\circ = \frac{71\pi}{90}$ or $0.7\overline{8}\pi$.

Examples - Radians to Degrees

Convert $\frac{\pi}{10}$ to degrees

$\frac{180^\circ}{\pi} * \frac{\pi}{10} = 18^\circ$

Convert $\frac{\pi}{2}$ to degrees

$\frac{180^\circ}{\pi} * \frac{\pi}{2} = 90^\circ$

Another way to convert radians to degrees is to simply replace the π with 180°:

$\frac{\pi}{10} = \frac{180^\circ}{10} = 18^\circ$

$\frac{\pi}{2} = \frac{180^\circ}{2} = 90^\circ$

Exercises 1.1

Convert each angle measure to decimal degrees.

1. $27^\circ\ 40'$	2. $38^\circ\ 20'$	3. $91^\circ\ 50'$
4. $34^\circ\ 10'$	5. $274^\circ\ 18'$	6. $165^\circ\ 48'$
7. $17^\circ\ 25'$	8. $63^\circ\ 35'$	9. $183^\circ\ 33'\ 36''$
10. $141^\circ\ 6'\ 9''$	11. $211^\circ\ 46'\ 48''$	12. $19^\circ\ 12'\ 18''$

Convert each angle measure to DMS notation.

13. 31.425°	14. 159.84°	15. 6.78°
16. 24.56°	17. 110.25°	18. 64.16°
19. 18.9°	20. 85.14°	21. 220.43°
22. 55.17°	23. 70.214°	24. 116.32°

Convert each angle measure from degrees to radians.

25. 30°	26. 120°	27. 45°
28. 225°	29. 60°	30. 150°
31. 90°	32. 270°	33. 15°
34. 36°	35. 12°	36. 104°

Convert each angle measure from radians to degrees.

37. $\frac{\pi}{4}$

38. $\frac{\pi}{5}$

39. $\frac{\pi}{3}$

40. $\frac{\pi}{6}$

41. $\frac{3\pi}{4}$

42. $\frac{7\pi}{3}$

43. $\frac{5\pi}{2}$

44. $\frac{7\pi}{4}$

45. $\frac{5\pi}{6}$

46. $\frac{2\pi}{3}$

47. π

48. $\frac{7\pi}{2}$

1.2 The Trigonometric Ratios

There are six common trigonometric ratios that relate the sides of a right triangle to the angles within the triangle. The three standard ratios are the sine, cosine and tangent. These are often abbreviated sin, cos and tan. The other three (cosecant, secant and cotangent) are the reciprocals of the sine, cosine and tangent and are often abbreviated csc, sec, and cot.

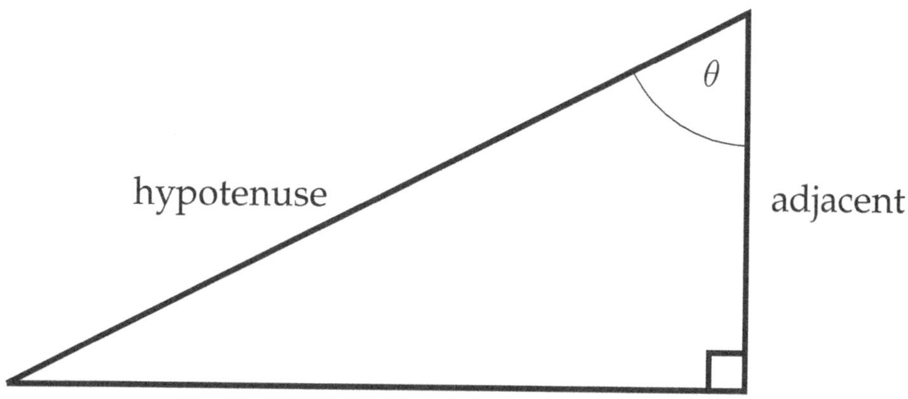

Given an angle situated in a right triangle, the sine function is defined as the ratio of the side opposite the angle to the hypotenuse, the cosine is defined as the ratio of the side adjacent to the angle to the hypotenuse and the tangent is defined as the ratio of the side opposite the angle to the side adjacent to the angle.

$$\sin\theta = \frac{opp}{hyp}$$

$$\cos\theta = \frac{adj}{hyp}$$

$$\tan\theta = \frac{opp}{adj}$$

A common mneumonic device to help remember these relationships is –SOHCAHTOA– which identifies the Sin as Opp over Hyp Cos as Adj over Hyp and the Tan as Opp over Adj.

An acute angle placed in the other position of a right triangle would have different oppposite and adjacent sides although the hypotenuse would remain the same.

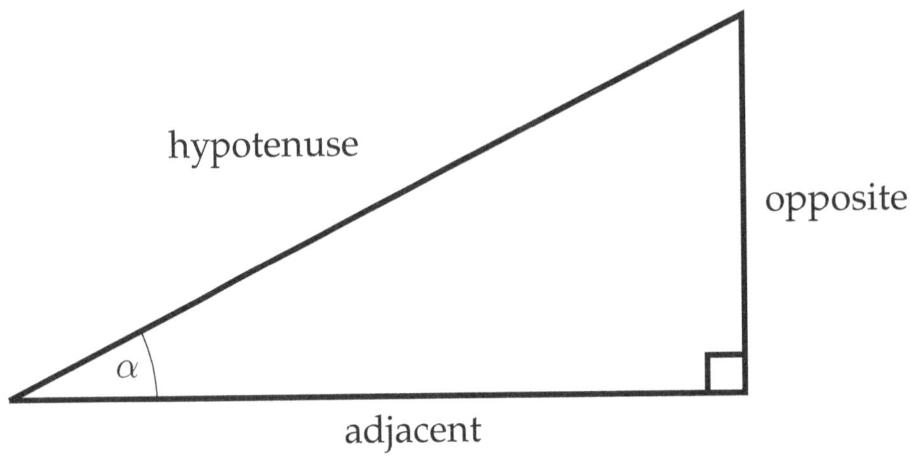

Examples: Trigonometric Ratios

Find $\sin\theta$, $\cos\theta$ and $\tan\theta$ for the given angle θ

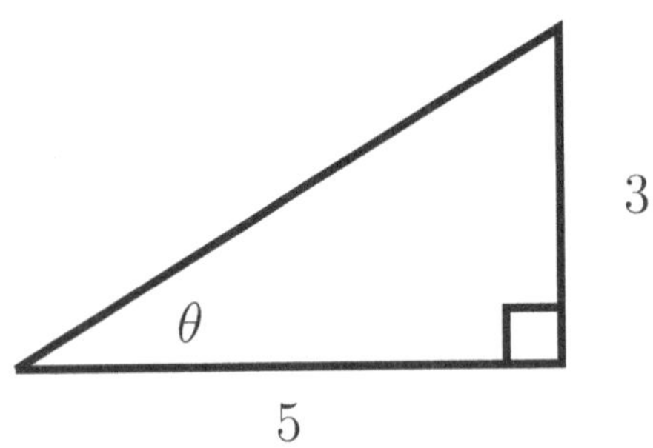

In order to find the sin and cos of the angle θ, we must first find the hypotenuse by using the Pythagorean Theorem ($a^2 + b^2 = c^2$).

Since we know the legs of the triangle, we can substitute these values for a and b in the Pythagorean Theorem:

$$3^2 + 5^2 = c^2$$

$$9 + 25 = c^2$$

$$34 = c^2$$

$$\sqrt{34} = c$$

Now that we know the hypotenuse ($\sqrt{34}$), we can determine the sin, cos and tan for the angle θ.

$$\sin\theta = \frac{3}{\sqrt{34}}$$

$$\cos\theta = \frac{5}{\sqrt{34}}$$

$$\tan\theta = \frac{3}{5}$$

Find $\sin\theta$**,** $\cos\theta$ **and** $\tan\theta$ **for the given angle** θ

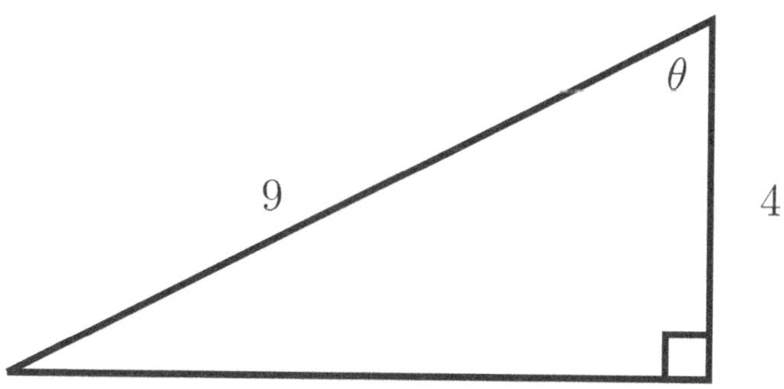

Again, in order to find the sin, cos and tan of the angle θ, we must find the missing side of the triangle by using the Pythagorean Theorem. Since, in this case, we know the hypotenuse and one of the legs, the value of the hypotenuse must be substituted for c and the length of the leg we're given can be substituted for either a or b.

$$4^2 + b^2 = 9^2$$

$$16 + b^2 = 81$$

$$b^2 = 65$$

$$b = \sqrt{65}$$

Now that we know the length of the other leg of the triangle ($\sqrt{65}$), we can determine the sin, cos and tan for the angle θ.

$$\sin\theta = \frac{\sqrt{65}}{9}$$

$$\cos\theta = \frac{4}{9}$$

$$\tan\theta = \frac{\sqrt{65}}{4}$$

In addition to the examples above, if we are given the value of one of the trigonometric ratios, we can find the value of the other two.

Example

Given that $\cos\theta = \frac{1}{3}$, find $\sin\theta$ and $\tan\theta$.

Given the information about the cosine of the angle θ, we can create a triangle that will allow us to find $\sin\theta$ and $\tan\theta$.

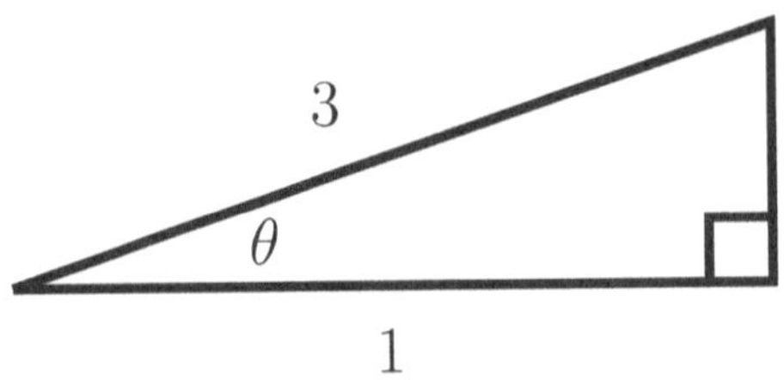

Using the Pythagorean Theorem, we can find the missing side of the triangle:

$$a^2 + 1^2 = 3^2$$

$$a^2 + 1 = 9$$

$$a^2 = 8$$

$$a = \sqrt{8} = 2\sqrt{2}$$

Then $\sin\theta = \frac{\sqrt{8}}{3}$ and $\tan\theta = \frac{\sqrt{8}}{1} = \sqrt{8}$.

You might say to yourself, "Wait a minute, just because the cosine of the angle θ is $\frac{1}{3}$, that doesn't necessarily mean that the sides of the triangle are 1 and 3, they could be 2 and 6, or 3 and 9 or any values n and $3n$."

This is true, and if the sides are expressed as n and $3n$, then the missing side would be $n\sqrt{8}$, so that whenever we find a trigonometric ratio, the n's will cancel out, so we just leave them out to begin with and call the sides 1 and 3.

Example

Given that $\tan\theta = \frac{\sqrt{5}}{7}$**, find** $\sin\theta$ **and** $\cos\theta$**.**

First we'll take the infomation about the tangent and use this to draw a triangle.

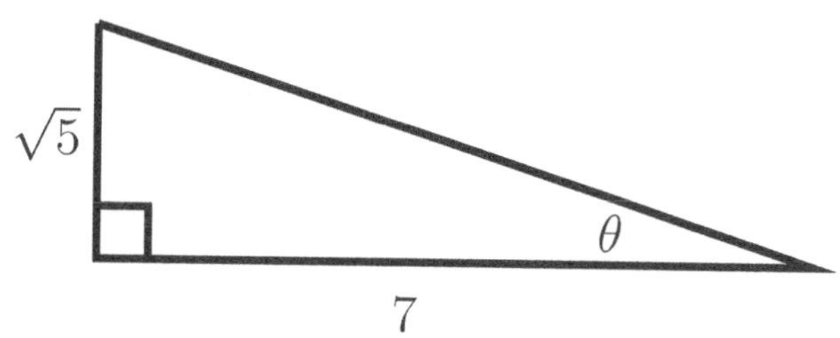

Then use the Pythagorean Theorem to find the missing side of the triangle:

$$\sqrt{5}^2 + 7^2 = c^2$$

$$5 + 49 = c^2$$

$$54 = c^2$$

$$\sqrt{54} = 3\sqrt{6} = c$$

So then:

$$\sin\theta = \frac{\sqrt{5}}{\sqrt{54}} = \sqrt{\frac{5}{54}}$$

$$\cos\theta = \frac{7}{\sqrt{54}} = \frac{7}{3\sqrt{6}}$$

Exercises 1.2

Find $\sin\theta$, $\cos\theta$ and $\tan\theta$ for the given triangles.

1. 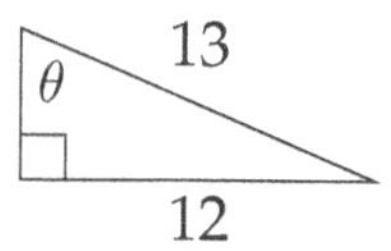

2. 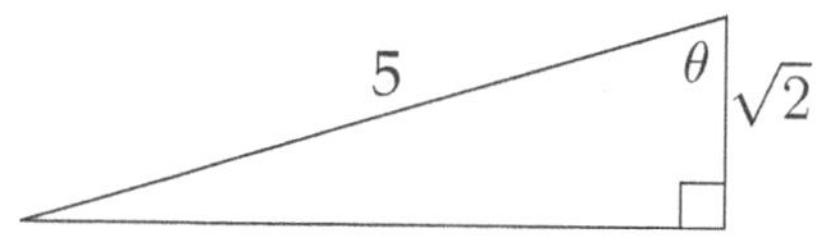

3. 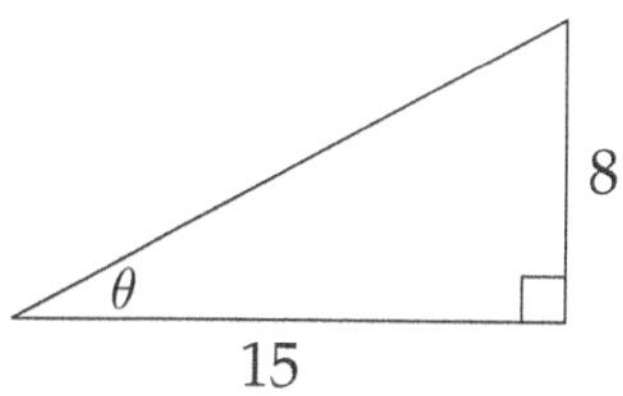

4. 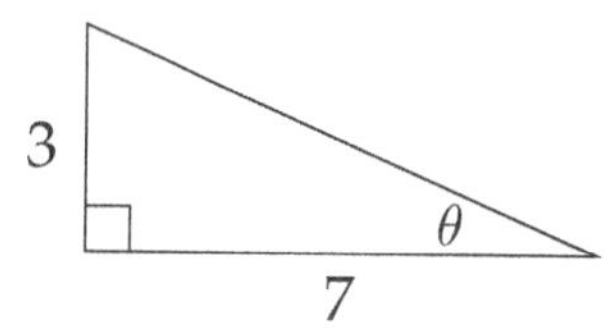

5. 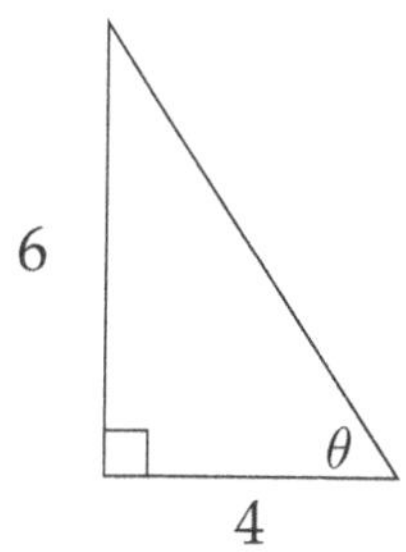

6. 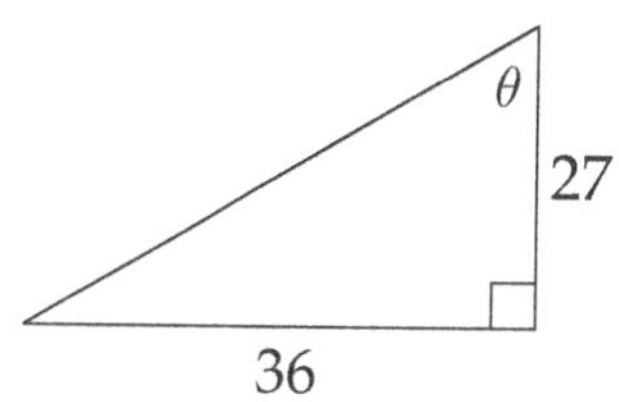

7. 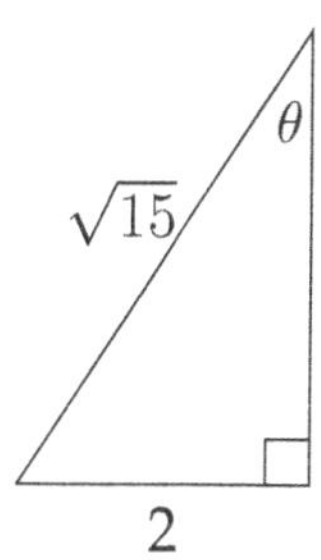

8. 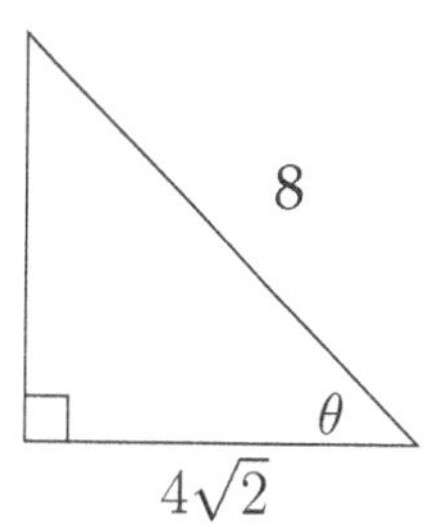

9. 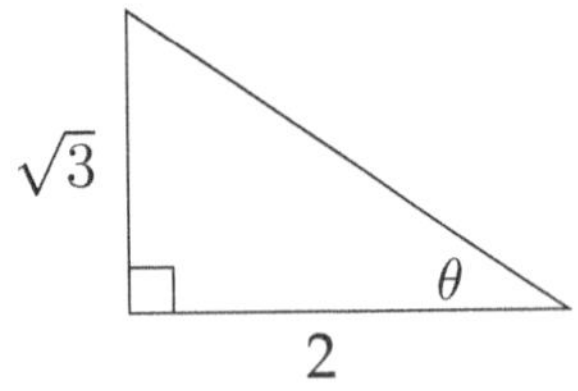

10. 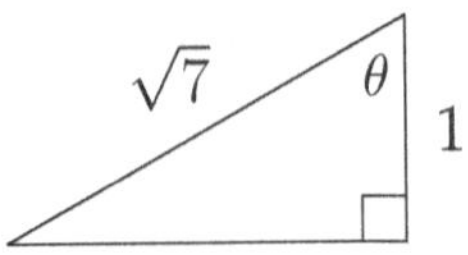

Use the information given to find the other two trigonometric ratios.

11. $\tan\theta = \frac{1}{2}$
12. $\sin\theta = \frac{3}{4}$
13. $\cos\theta = \frac{3}{\sqrt{20}}$
14. $\tan\theta = 2$
15. $\sin\theta = \frac{5}{\sqrt{40}}$
16. $\sin\theta = \frac{7}{10}$
17. $\cos\theta = \frac{9}{40}$
18. $\tan\theta = \sqrt{3}$
19. $\cos\theta = \frac{1}{2}$
20. $\cos\theta = \frac{3}{7}$
21. $\sin\theta = \frac{\sqrt{5}}{7}$
22. $\tan\theta = 1.5$

1.3 Solving Triangles

Using information about some of the sides and angles of a triangle in order to find the unknown sides and angles is called "solving the triangle." If two sides of a triangle are known, the Pythagorean Theorem can be used to find the third side. If one of the acute angles in a right triangle is known, the other angle will be its complement with their sum being 90°.

Suppose that we have a right triangle in which we know the sides, but no angles. Another situation could involve knowing the angles but just one side. How could we solve for the missing measurements in these situations?

Solving problems like these uses precalculated values of the trigonometric ratios to match the lengths with the appropriate angles and vice versa. Up until the 1980's, these values were printed in tables that were included in the back of every textbook (along with tables of logarithms), but have recently been programmed into calculators using methods that are studied in Calculus.

Most calculators have a button or function designed to find the inverse sine, inverse cosine and inverse tangent ($\sin^{-1}$, $\cos^{-1}$, and $\tan^{-1}$), these are the functions that tell you the measure of the angle that has a sine, cosine or tangent equal to a particular value.

For example, if we are given an angle θ and know that the $\sin\theta = \frac{1}{2}$:

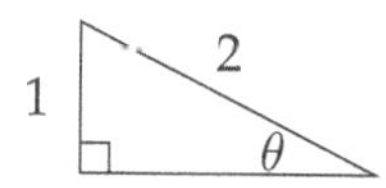

Then we can find $\sin^{-1}(\frac{1}{2})$ on a calculator, which should return a value of 30°. If the calculator is in radian mode, it will return a value of ≈ 0.523598776. If you divided this number by π, you would get $0.1\overline{6}$, which means that $0.523598776 \approx \frac{\pi}{6}$. In this chapter we will work mainly in degrees. In Chapter 2, when we graph the trigonometric functions we will typically use radian measure.

Example 1

Solve the triangle. Round side lengths to the nearest 100th and angles to the nearest 10th of a degree.

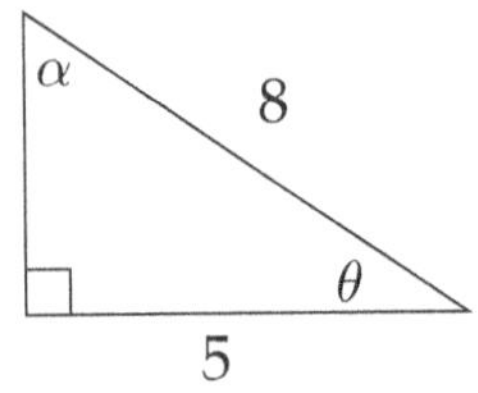

We can find the third side of the triangle by using the Pythagorean Theorem.

$$a^2 + 5^2 = 8^2$$
$$a^2 + 25 = 64$$
$$a^2 = 39$$
$$a = \sqrt{39} \approx 6.24$$

When solving problems of this type, I encourage people to use the most accurate values that are available in the problem. This way, there is less chance for rounding error to occur.

If we take the values for the sides that were given in the problem (5 and 8), then we can say that

$$\cos\theta = \frac{5}{8}$$
$$\theta = \cos^{-1}\left(\frac{5}{8}\right)$$
$$\theta \approx 51.3^\circ$$

Then α would be $90^\circ - 51.3^\circ = 38.7^\circ \approx \alpha$.

Example 2

Solve the triangle. Round side lengths to the nearest 100th and angles to the nearest 10th of a degree.

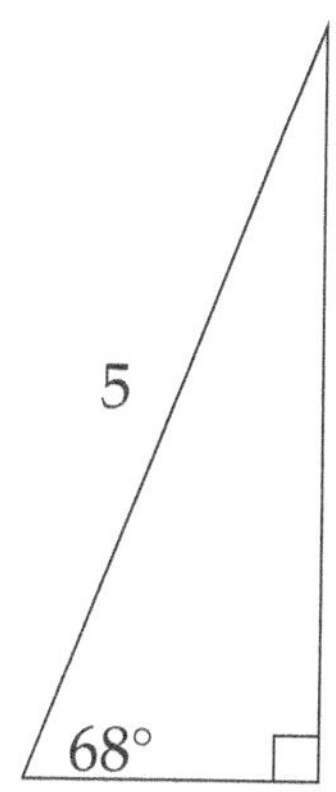

First, we can find the other angle in the right triangle: $90° - 68° = 22°$.

Next, to find the sides, we choose a trigonometric ratio for which we know one of the sides. In this problem, we can use either the sine or the cosine.

$$\sin 68° = \frac{a}{5}$$

Approximating $\sin 68°$ on a calculator:

$$0.9272 \approx \frac{a}{5}$$

$$5 * 0.9272 \approx a$$

$$4.6 \approx a$$

When approximating a trigonometric value from the calculator, it is important to use at least 4 decimal places of accuracy. Again, this is to avoid rounding errors.

To solve for the remaining side we can either use the Pythagorean Theorem or use the method demonstrated above, but with the $\cos 68°$.

$$\cos 68^\circ = \frac{b}{5}$$

Approximating $\cos 68^\circ$ on a calculator:

$$0.3746 \approx \frac{b}{5}$$
$$5 * 0.3746 \approx b$$
$$1.9 \approx b$$

If we use the Pythagorean Theorem with two sides of the triangle to find the third, then we would say that:

$$b^2 + 4.6^2 = 5^2$$
$$b^2 + 21.16 = 25$$
$$b^2 = 3.84$$
$$b = \sqrt{3.84} \approx 1.959 \approx 2.0$$

The rounding error in this example comes from the fact that the first side we found was not exactly 4.6. If we wanted a more accurate answer that matches the answer we found using the cosine ratio, we just need more accuracy in the leg of the triangle we found.

Calculating $5 * \sin 68^\circ \approx 4.636$ should provide enough accuracy.

$$b^2 + 4.636^2 = 5^2$$
$$b^2 + 21.492496 = 25$$
$$b^2 = 3.507504$$
$$b = \sqrt{3.507504} \approx 1.8728 \approx 1.9$$

Example 3

Solve the triangle. Round side lengths to the nearest 100th and angles to the nearest 10th of a degree.

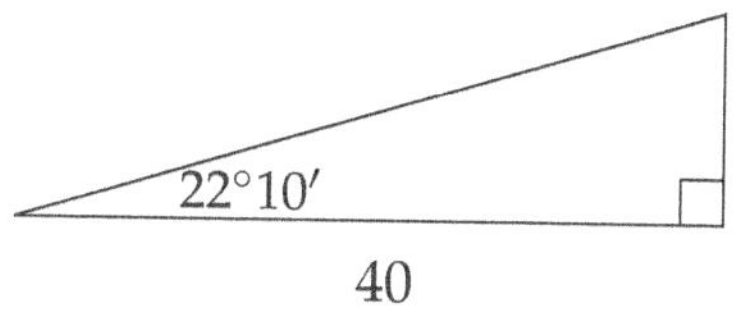

If we convert the angle $22°10'$ to $22.1\overline{6}^{\circ}$, then the other acute angle in the right triangle is $90° - 22.1\overline{6}^{\circ} = 67.8\overline{3}^{\circ}$ or $67°50'$.

Finding the remaining sides requires the use of either the cosine or tangent function.

$$\cos 22.1\overline{6}^{\circ} = \frac{40}{c}$$

Approximating $\cos 22.1\overline{6}^{\circ}$ on a calculator:

$$0.9261 \approx \frac{40}{c}$$

Next we need to multiply on both sides by c:

$$c * 0.9261 \approx \frac{40}{c} * c$$

$$0.9261c \approx 40$$

Then divide on both sides by 0.9261:

$$\frac{0.9261c}{0.9261} \approx \frac{40}{0.9261}$$

$$c \approx 43.2$$

To find the other leg of the triangle, we can use the tangent ratio.

$$\tan 22.1\overline{6}^\circ = \frac{a}{40}$$

Approximating $\tan 22.1\overline{6}^\circ$ on a calculator:

$$0.4074 \approx \frac{a}{40}$$

Then, multiply on both sides by 40:

$$40 * 0.4074 \approx a$$

$$16.3 \approx a$$

We can check this answer by the Pythagorean Theorem:

$$16.3^2 + 40^2 = 1865.69$$

$$\sqrt{1865.69} \approx 43.2$$

Exercises 1.3

In each problem below, solve the triangle. Round side lengths to the nearest 100th and angle measures to the nearest 10th of a degree.

1. 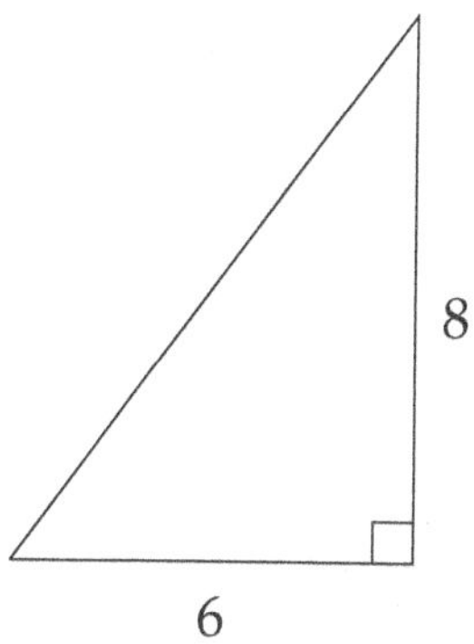

2. 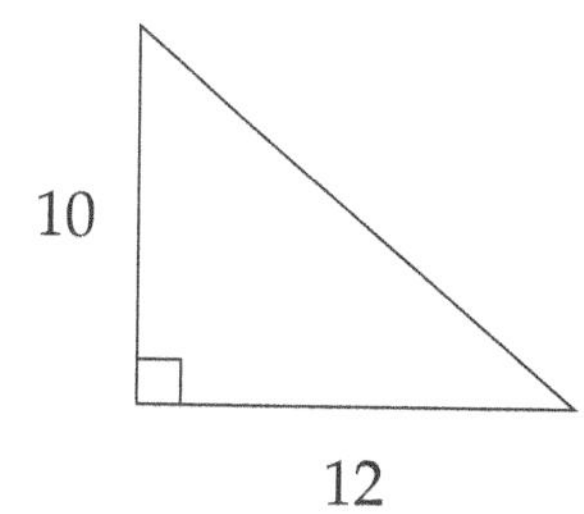

3. 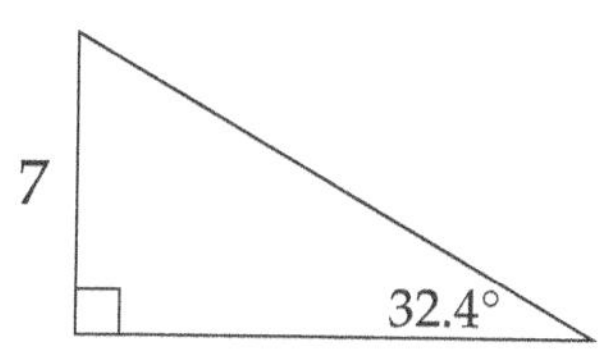

4. 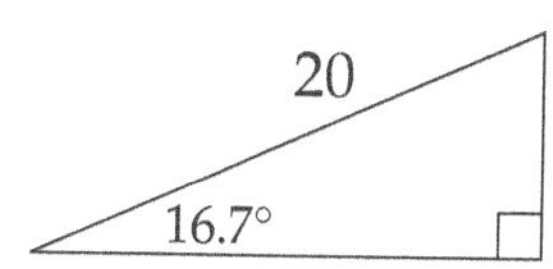

5. 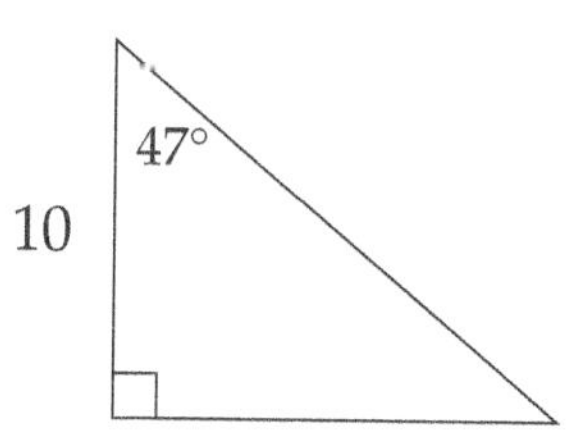

6. 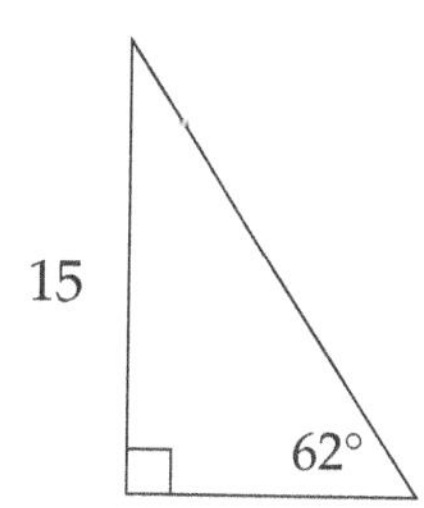

7. 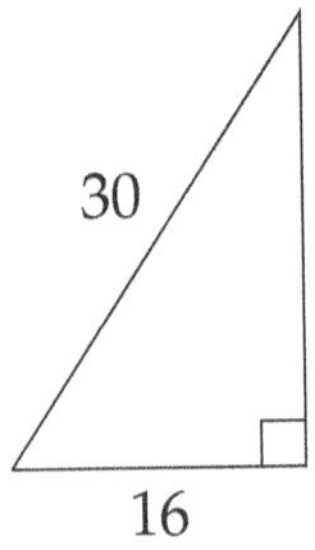

8. 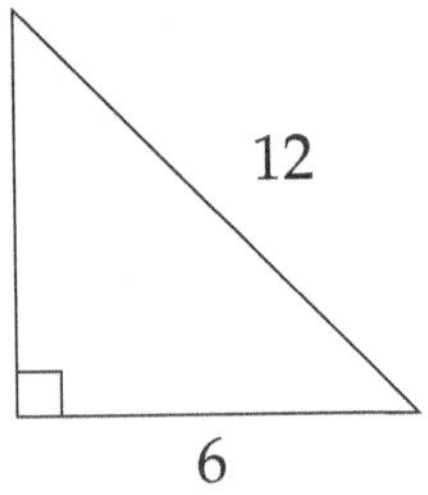

9. 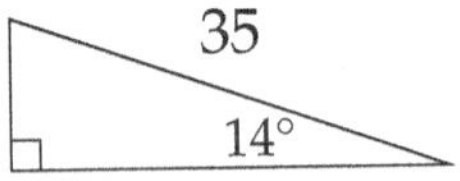

10. 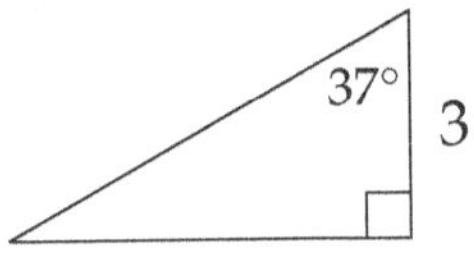

11. 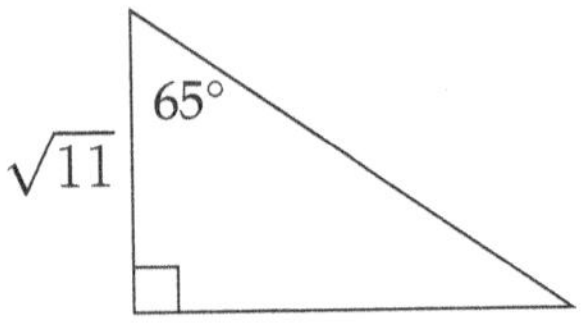

12. 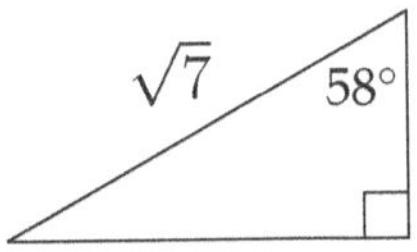

1.4 Applications

Trigonometry is often used for what is called "indirect measurement." This is a method of measuring inaccessible distances by using the relationships between lengths and angles within a triangle. Two simple examples of this process are measuring the height of a tall tree and measuring the distance across a body of water. In both cases, while it might be possible to measure the distance directly, it is often much easier to use indirect measurement.

In one example of indirect measurement, the angle of elevation of an object can be used to create a right triangle in which one angle and one side are known. The other sides of the triangle may then be solved for. In the problems in this text, the angle of elevation will typically be given in the problem. In order to actually measure the angle of elevation of an object, it is possible to use a simple protractor.

If you wanted to measure the height of a tall tree that sits on flat ground, you could use a specially modified protractor to do this. Modifying the protractor by tying a weight to the end of a string and tying the other end of the string through the hole in the protractor will help to measure the angle of elevation. Once the protractor is ready, hold it upside down and sight the top of the tree along the straight edge of the protractor. The weight hanging down will show the complement to the angle of elevation. In other words, if the angle of elevation is 20°, the string will mark out a measurement of 70°on the protractor.

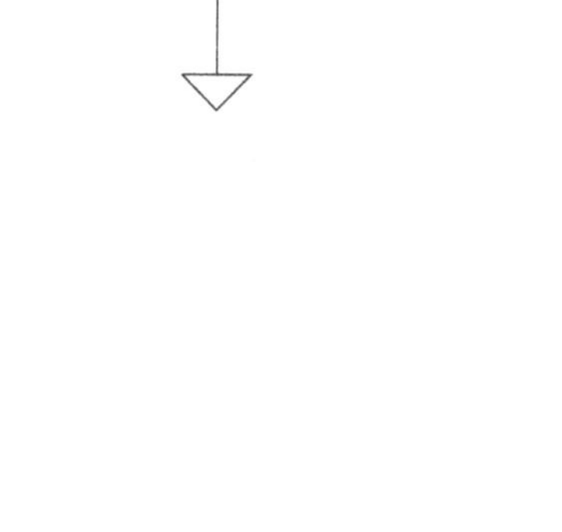

Closely related to the concept of the angle of elevation is the angle of depression. This is the angle that is formed by looking down on something from above.

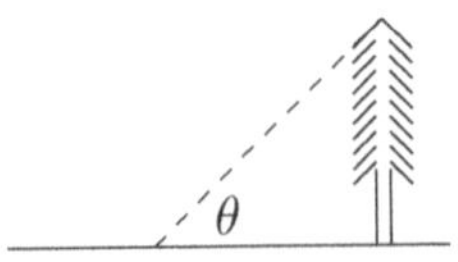

The angle of elevation is θ.

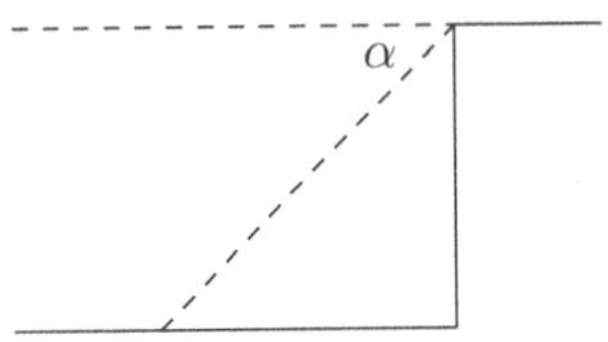

The angle of depression is α.

In a situation in which the angle of depression is measured, the angle of elevation and the angle of depression are alternate interior angles, which makes them equal.

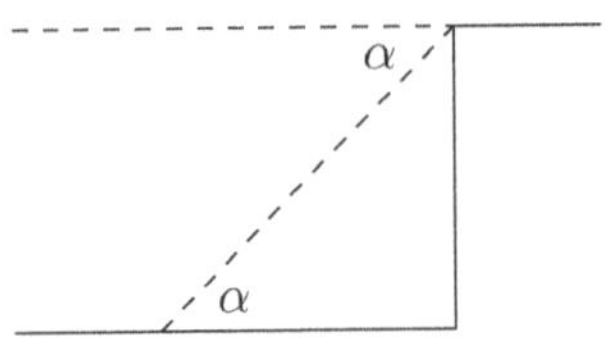

The angle of elevation equals the corrresponding angle of depression.

Example 1

Pacing off 100 ft. on flat ground from the base of a tree, a forester measures the angle of elevation to the top of the tree as 65°. What is the height of the tree?

The situation described in the problem creates a diagram like the one below:

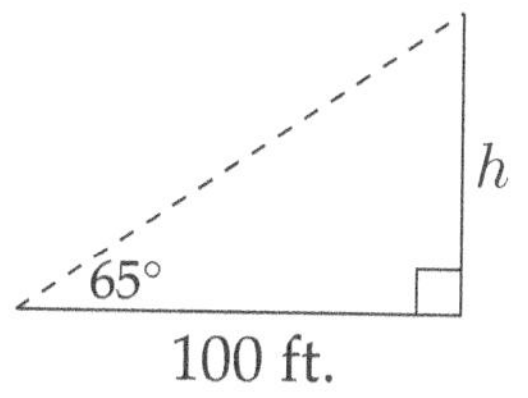

Since this is a right triangle, we can use an appropriate trigonometric ratio to find the height of the tree. In this case,

$$\tan 65^\circ = \frac{h}{100}.$$

$$100 * \tan 65^\circ = h$$

$$214.45 \approx h$$

So, the tree is about 214.45 feet tall.

Example 2

From the top of a building 125 feet tall, the angle of depression of an intersection is 34°. How far from the base of the building is the intersection?

As in the previous example, it is often helpful to draw a diagram.

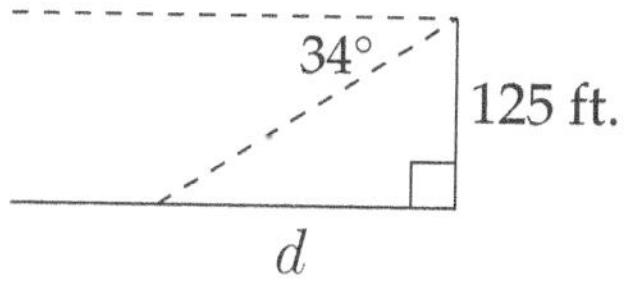

Again, the angle of elevation will be equal to the corresponding angle of depression, so we can use the triangle as seen below to solve the problem:

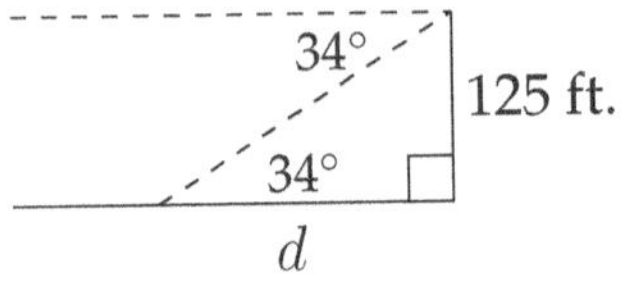

In this problem

$$\tan 34^\circ = \frac{125}{d}$$

Multiply on both sides by d

$$d * \tan 34^\circ = 125$$

Then divide on both sides by $\tan 34^\circ$

$$\frac{d \tan 34^\circ}{\tan 34^\circ} = \frac{125}{\tan 34^\circ}$$

$$d \approx 185.32$$

Example 3

Sometimes a problem involves both an angle of elevation and an angle of depression.

From the roof of a house 20 feet off the ground, the angle of elevation of the top of an apartment building is 63°and the angle of depression to the base of the building is 24°. How far away from the house is the apartment building? How tall is the apartment building?

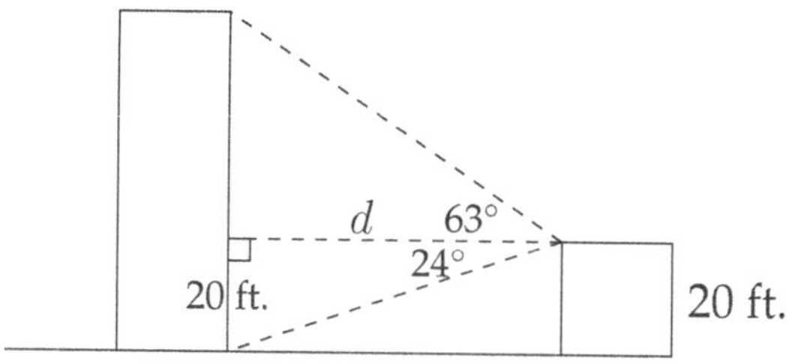

If we work on the bottom triangle first, then we know that the height of the triangle is 20 ft. and the angle opposite this side is 24°. So, we can say that:

$$\tan 24^\circ = \frac{20}{d}$$

$$d = \frac{20}{\tan 24^\circ}$$

$$d \approx 45 ft.$$

Now that we know that the apartment building is 45 feet away, we can use the upper triangle to determine the height ofthe building.

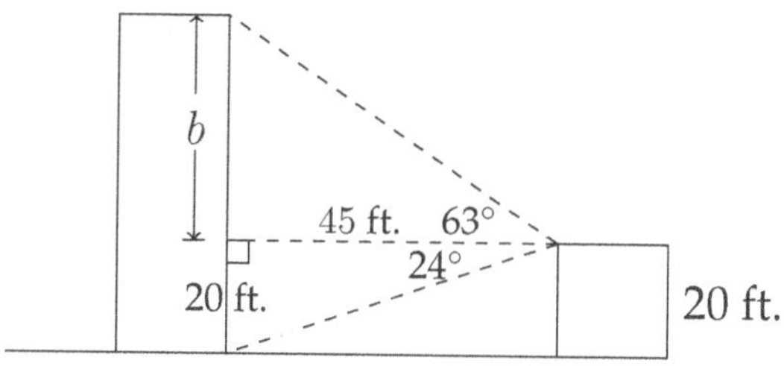

$$\tan 63^\circ = \frac{b}{45}$$

$$45 * \tan 63^\circ = b$$

$$88.3 ft. \approx b$$

$$20 + 88.3 = 108.3 ft.$$

Since the variable b only represents the part of the building that is in the second triangle, we need to add 20 feet to b to find the actual height of the building.

Problems on bearing and direction

Some applications of trigonometry involve ship navigation. One common method used to describe direction in this type of problem is what is known as bearing. Bearing describes a direction by the angle deviation from north or south. For example, the direction we typically describe as northeast is exactly halfway between north and east. The bearing for northeast would be $N45°E$, and is read as "Forty-five degrees east of north."

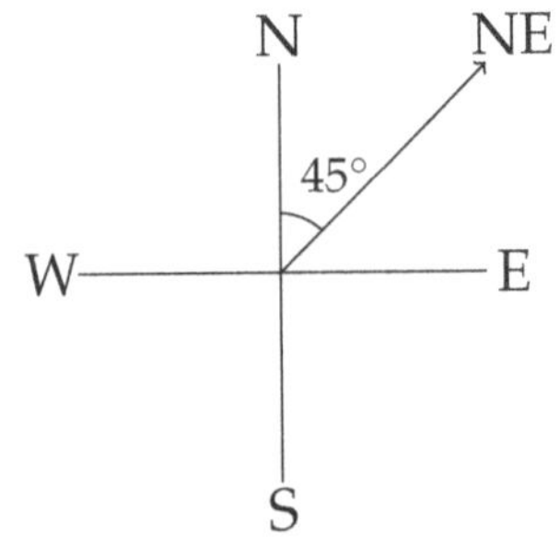

Forty-five degrees East of North: N45°E

Here are a few examples of what a bearing looks like in a N-S-E-W diagram.

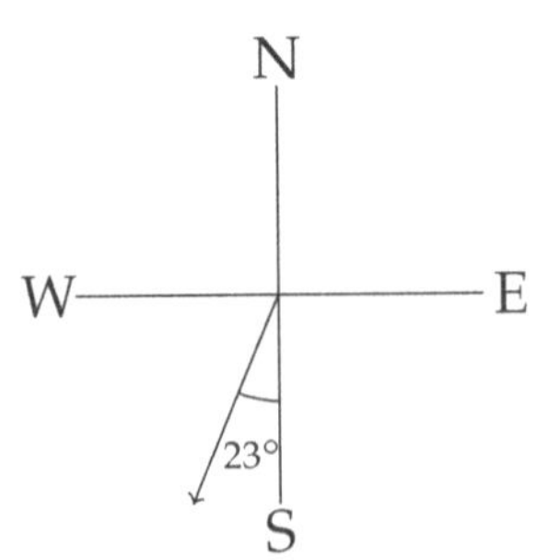

Twenty-three degrees West of South: S23°W

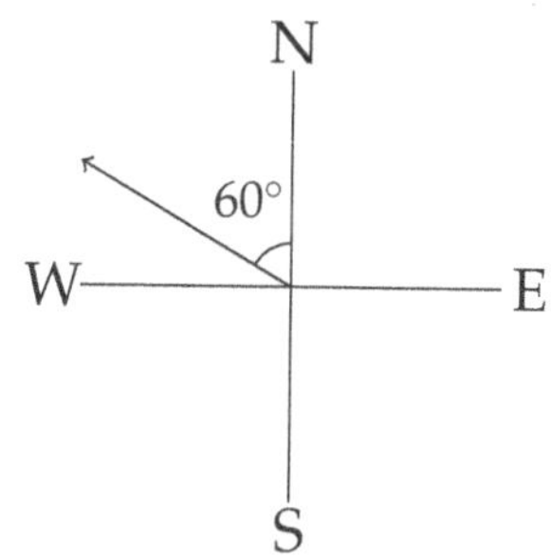

Sixty degrees West of North: N60°W

Example 4

Santa Rosa, California is 7 miles due north of Rohnert Park. Bodega Bay is 19 miles due west of Rohnert Park (as the crow flies). What is the bearing of Santa Rosa from Bodega Bay?

First, it would be helpful to draw a diagram to represent the situation:

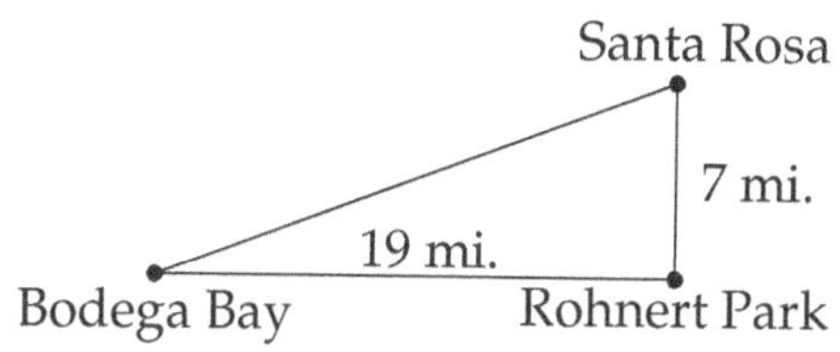

To answer the question we'll need another diagram:

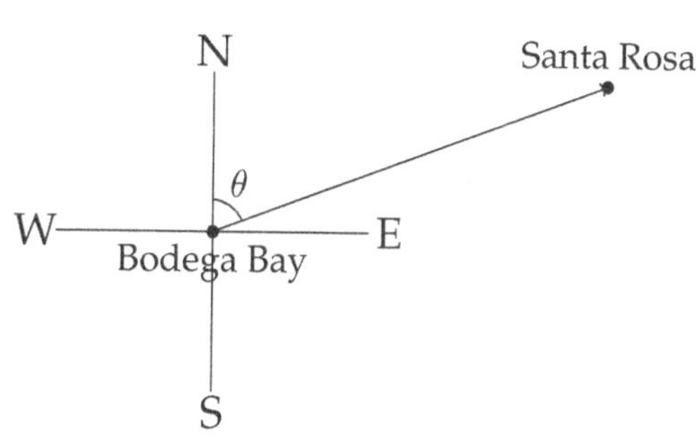

If we knew the angle θ, then we could conclude that the bearing of Santa Rosa from Bodega Bay is θ degrees East of North. From the previous diagram:

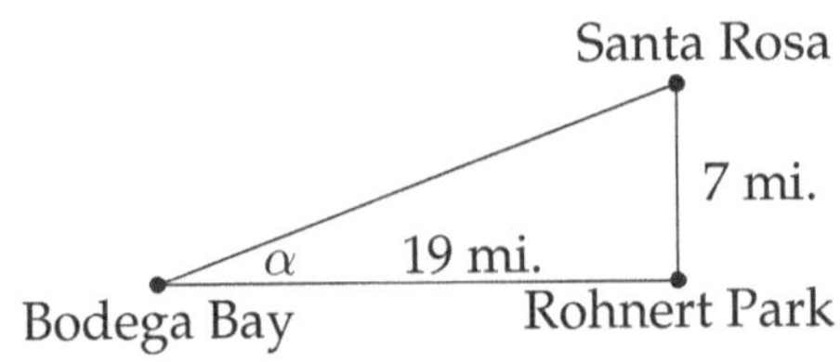

we can see that we can't find θ directly, but we can find the complement of θ.

$$\tan\alpha = \frac{7}{19}$$

$$\alpha \approx 20.2^\circ$$

$$\text{Therefore, } \theta \approx 90^\circ - 20.2^\circ \approx 69.8^\circ$$

That means that the bearing of Santa Rosa from Bodega Bay is $N69.8^\circ E$, or 69.8°East of North.

Exercises 1.4

Round answers to the nearest 10th.

1. From the top of a lighthouse 180 feet above sea level, the angle of depression to a ship in the ocean is 28°. How far is the ship from the base of the lighthouse?

2. A helicopter that is 700 feet in the air measures the angle of depression to a landing pad as 24°. How far is the landing pad from the point directly beneath the helicopter's current position?

3. An 88 foot tree casts a shadow that is 135 feet long. What is the angle of elevation of the sun?

4. A 275 foot guy wire is attached to the top of a communication tower. If the wire makes an angle of 53°with the ground, how tall is the tower?

5. A woman standing on a hill sees a building that she knows is 55 feet tall. The angle of depression to the bottom of the building is 27°and the angle of elevation to the top of the building is 35°. Find the straight line distance from the woman to the building.

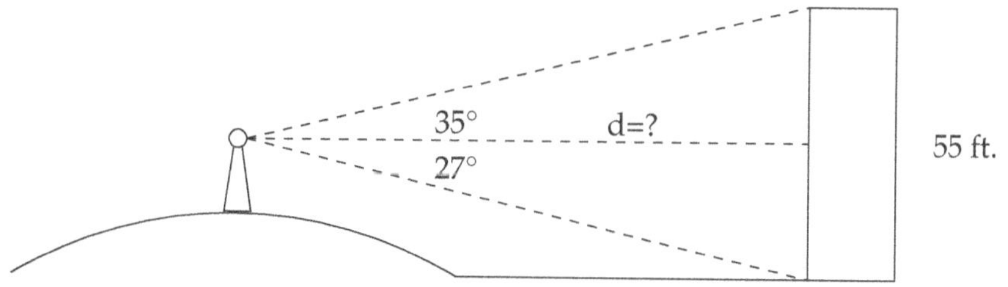

6. To measure the height of the cloud cover at an airport, a spotlight is shined upward at an angle of 62°. An observer 1000 feet away measures the angle of elevation to the spotlight to be 36°. Find the height of the cloud cover.

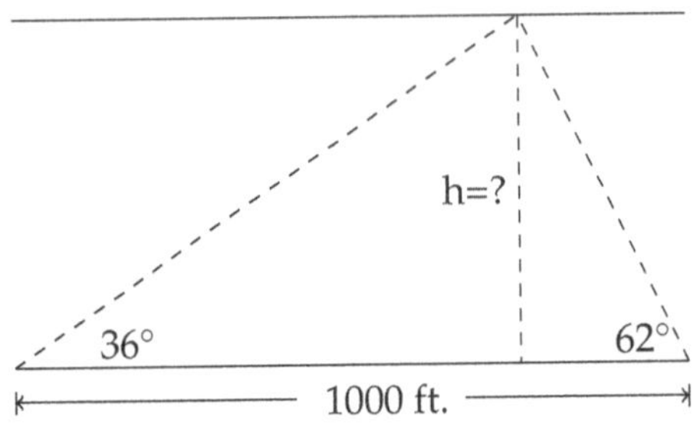

7. At ground level, a water tower is 430 feet from the base of a building. From one of the upper floors of the building, the angle of elevation to the top of the water tower is 15°and the angle of depression to the bottom of the water tower is 28°. How tall is the water tower? How high off the ground is the observer?

8. A small airplane is flying at the altitude of 7000 feet following a straight road directly beneath it. A car in front of the airplane is sighted with an angle of depression of 72°and a car behind the plane is sighted with an angle of depression of 48°. How far apart are the cars?

9. From a point on the floor, the angle of elevation to the top of a doorway is 43°. The angle of elevation to the ceiling directly above the doorway is 56°. If the ceiling is 10 feet above the floor, how high is the doorway? How far in front of the doorway were the angles of elevation measured?

10. A man standing on the roof of a building 70 feet high looks at the building next door. The angle of depression to the roof of the building next door is 36°. The angle of depression to the bottom of the building next door is 65°. How tall is the building next door?

11. A boat leaves the harbor and travels 30 miles in the direction of $N38°W$. The boat turns 90°and then travels in the direction $S52°W$ for 12 miles. At that time, how far is the boat from the harbor and what is the bearing of the boat from the harbor entrance?

12. A man walking in the desert travels 1.6 miles in the direction $S57°E$. He then turns 90°and continues walking for 3.2 miles in the direction $N33°E$. At that time, how far is he from his starting point and what is his bearing from the starting point?

13. Madras, Oregon is 26 miles due north of Redmond. Prineville is due east of Redmond and $S34°42'E$ from Madras. How far is Prineville from Redmond?

14. Raymond, Washington is 22 miles due south of Aberdeen. Montesano is due east of Aberdeen and $N26°34'E$ from Raymond. How far is Montesano from Raymond?

15. A boat travels on a course bearing of $S41°40'W$ for 84 miles. How far south and how far west is the boat from its starting point?

16. A boat travels on a course bearing of $N17°10'E$ for 10 miles. How far north and how far east is the boat from its starting point?

1.5 More Applications

Sometimes solving problems involving right triangles requires the use of a system of equations. A common method for determining the height of an object whose base is inaccessible is that of measuring the angle of elevation from two different places in front of the object. If you measure the angle of elevation to the top of of a radio antenna as 74°, then walk back 50 feet and measure the angle of elevation to the top of the antenna as 61°, then we would have something like the diagram below:

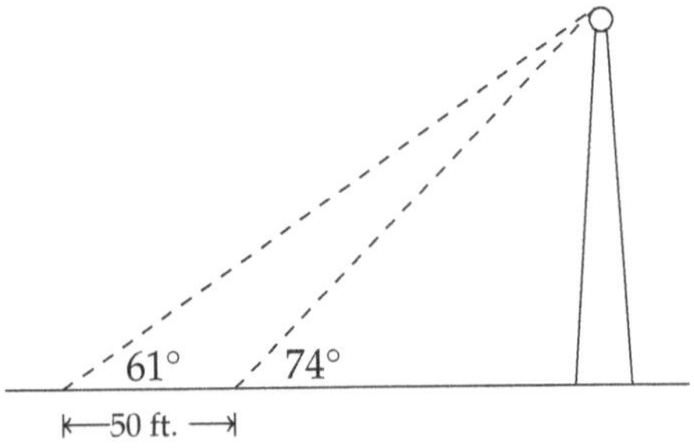

One of the first things we can do is introduce some labels for the unknown distances:

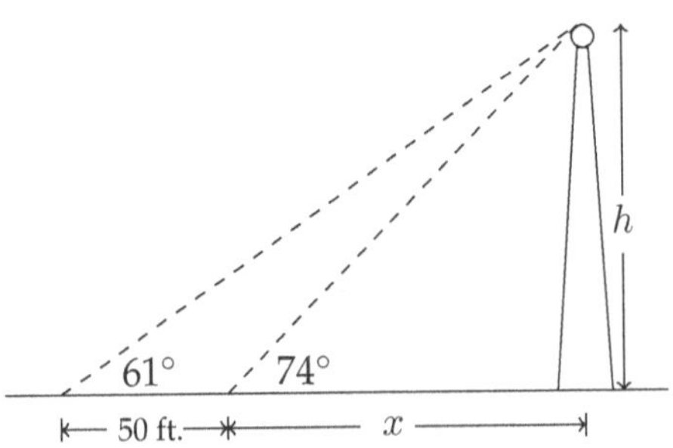

Then, we can say that:

$$\tan 74^\circ = \frac{h}{x}$$

$$\tan 61^\circ = \frac{h}{x+50}$$

To solve this system of equations, we'll set the first one equal to h:

$$\tan 74^\circ = \frac{h}{x}$$

$$x * \tan 74^\circ = h$$

Then, substitute this into the second equation:

$$\tan 61^\circ = \frac{h}{x+50}$$

$$\tan 61^\circ = \frac{x \tan 74^\circ}{x+50}$$

Multiply on both sides by $x + 50$:

$$(x+50)\tan 61^\circ = \frac{x \tan 74^\circ}{x+50}(x+50)$$

So,

$$(x+50)\tan 61^\circ = x \tan 74^\circ$$

There are two options to solve this equation - we can hold on to the tangents as they are and solve for x in terms $\tan 74^\circ$ and $\tan 61^\circ$, or we can approximate $\tan 74^\circ$ and $\tan 61^\circ$ and generate an approximate value for x and h. First we'll approximate:

$$(x+50)\tan 61^\circ = x \tan 74^\circ$$

$$(x+50) * 1.804 \approx 3.4874x$$

$$1.804x + 90.2024 \approx 3.4874x$$

$$90.2024 \approx 1.6834x$$

$$53.58 \approx x$$

$$x * \tan 74^\circ = h$$

$$53.58 * \tan 74^\circ \approx h$$

$$186.87 \text{ feet} \approx h$$

The other method is a little tricky algebraically:

$$(x + 50)\tan 61^\circ = x \tan 74^\circ$$

$$x \tan 61^\circ + 50 \tan 61^\circ = x \tan 74^\circ$$

$$50 \tan 61^\circ = x \tan 74^\circ - x \tan 61^\circ$$

$$50 \tan 61^\circ = x(\tan 74^\circ - \tan 61^\circ)$$

$$\frac{50 \tan 61^\circ}{(\tan 74^\circ - \tan 61^\circ)} = x$$

At this point, you can approximate the value of x and solve for h, or express the value of h exactly as

$$\tan 74^\circ * \frac{50 \tan 61^\circ}{(\tan 74^\circ - \tan 61^\circ)} = h$$

Exercises 1.5

1. Find the indicated height h.

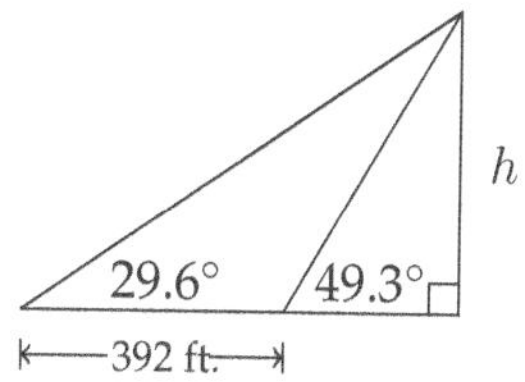

2. Find the indicated height h.

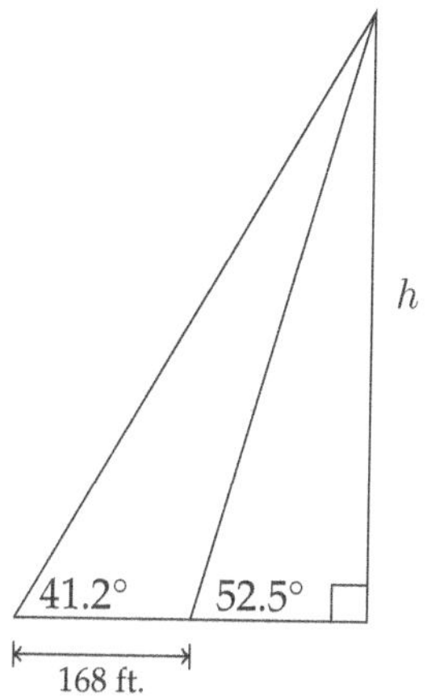

3. A small airplane flying at an altitude of 5300 feet sights two cars in front of the plane traveling on a road directly beneath it. The angle of depression to the nearest car is 62°and the angle of depression to the more distant car is 41°. How far apart are the cars?

4. A hot air balloon is flying above a straight road. In order to estimate their altitude, the people in the balloon measure the angles of depression to two consecutive mile markers on the same side of the balloon. The angle to the closer marker is 17°and the angle to the farther one is 13°. At what altitude is the balloon flying?

5. To estimate the height of a mountain, the angle of elevation from a spot on level ground to the top of the mountain is measured to be 32°. From a point

1000 feet closer to the mountain, the angle of elevation is measured to be 35°. How high is the mountain above the ground from which the measurements were taken?

6. The angle of elevation from a point on the ground to the top of a pyramid is 35°30′. The angle of elevation from a point 135 feet farther back to the top of the pyramid is 21°10′. What is the height of the pyramid?

7. An observer in a lighthouse 70 feet above sea level sights the angle of depression of an approaching ship to be 15°50′. A few minutes later the angle of depression is sighted at 35°40′. Find the distance traveled by the ship during that time.

8. To estimate the height of a tree, one forester stands due west of the tree and another forester stands due north of the tree. The two foresters are the same distance from the base of the tree and they are 45 feet from each other. If the angle of elevation for each forester is 40°, how tall is the tree?

9. A ship is anchored off of a long straight shoreline that runs east to west. From two observation points located 10 miles apart on the shoreline, the bearings of the ship from each observation point are $S35°E$ and $S17°W$. How far from shore is the ship?

10. From fire lookout Station Alpha the bearing of a forest fire is $N52°E$. From lookout Station Beta, sited 6 miles due east of Station Alpha, the bearing is $N38°W$. How far is the fire from Station Alpha?

11. From a point 200 feet from the base of a church, the angle of elevation to the top of the steeple is 28°, while the angle of elevation to the bottom of the steeple is 20°. How high off the ground is the top of the steeple?

12. A television tower 75 feet tall is installed on the top of a building. From a point on the ground in front of the building, the angle of elevation to the top of the tower is 62°and the the angle of elevation to the bottom of the tower is 44°. How tall is the building?

Chapter 2

Graphing the Trigonometric Functions

2.1 Trigonometric Functions of Non-Acute Angles

In Chapter 1, we learned about the trigonometric functions of positive acute angles that occur within right triangles. If we wish to extend the definition of the trigonometric functions, then we need to define how to determine the values for the sine and cosine of other angles. To do this, consider a right triangle drawn on the coordinate axes. The positive acute angle θ will be the angle created between the x-axis and the hypotenuse of the triangle. The lengths of the two legs of the triangle will be the x and y coordinates of a point in the first quadrant.

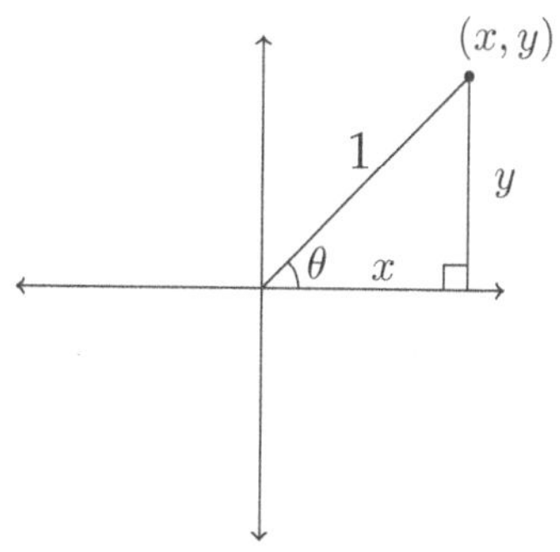

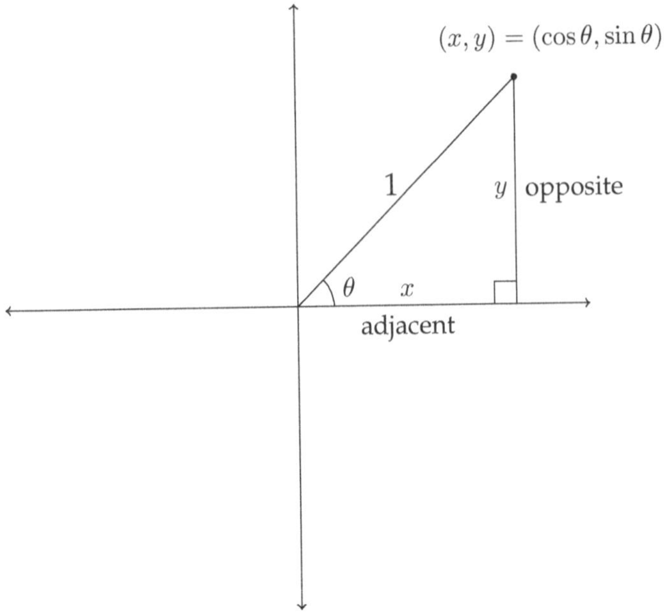

In the picture above we see a triangle in the first quadrant with a hypotenuse of 1. In this situation, the value of $\sin\theta = \frac{opp}{hyp} = \frac{y}{1} = y$, which is just the y-coordinate of the point at the top of the triangle. Correspondingly, the value of $\cos\theta = \frac{adj}{hyp} = \frac{x}{1} = x$, or the value of the x-coordinate of the same point.

This allows us to find the sine or cosine for what are known as the quadrantal angles - the angles that are multiples of 90°. If we look at the unit circle (the circle with a radius of 1), then we can see the values of the sine and cosine for these angles.

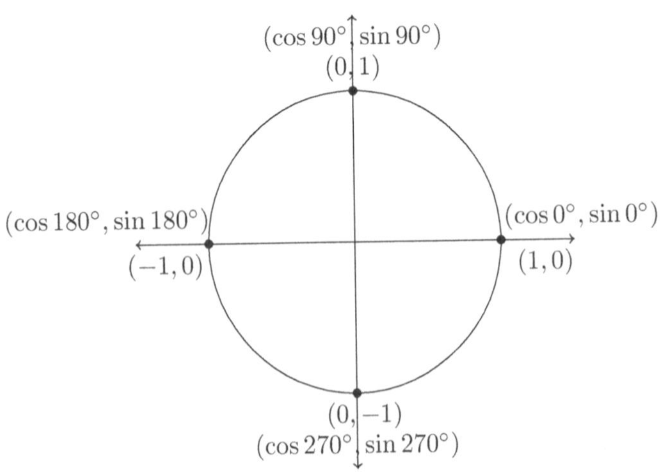

In the previous diagram, we see the values for the sine and cosine of the quadrantal angles:

$\cos 0° = 1$	$\cos 90° = 0$	$\cos 180° = -1$	$\cos 270° = 0$
$\sin 0° = 0$	$\sin 90° = 1$	$\sin 180° = 0$	$\sin 270° = -1$

If we take a radius of length 1 and rotate it counter-clockwise in the coordinate plane, the x and y coordinates of the point at the tip will correspond to the values of the cosine and sine of the angle that is created in the rotation. Let's look at an example in the second quadrant. If we rotate a line segment of length 1 by 120°, it will terminate in Quadrant II.

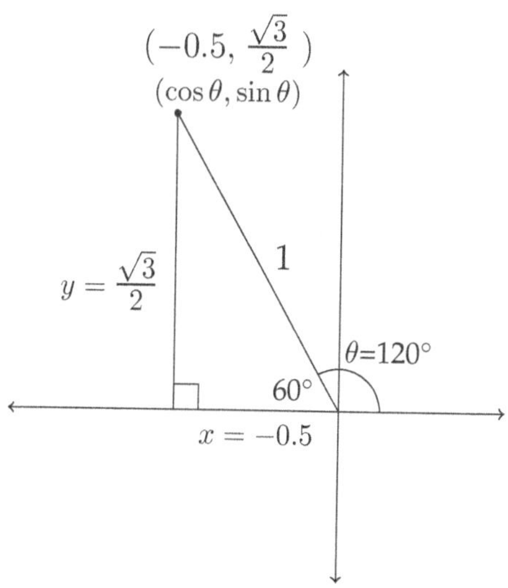

In the diagram above we notice several things. The radius of length 1 has been rotated by 120°into Quadrant II. If we then drop a perpendicular line from the endpoint of the radius to the x-axis, we create a triangle in Quadrant II. Notice that the angle supplementary to 120°appears in the triangle and this allows us to find the lengths of the sides of the triangle and hence the values for the x and y coordinates of the point at the tip of the radius.

Whenever an angle greater than 90°is created on the coordinate axes, simply drop a perpendicular to the x-axis. The angle created is the reference angle. The values of the trigonometric functions of the angle of rotation and the reference angle will differ only in their sign $(+, -)$. On the next page are examples for Quadrants II, III, and IV.

Quadrant II

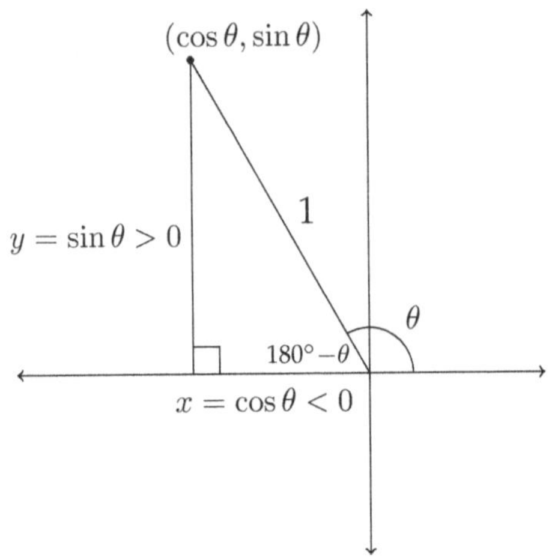

In Quadrant II, the cosine is negative and the sine is positive.

Quadrant III

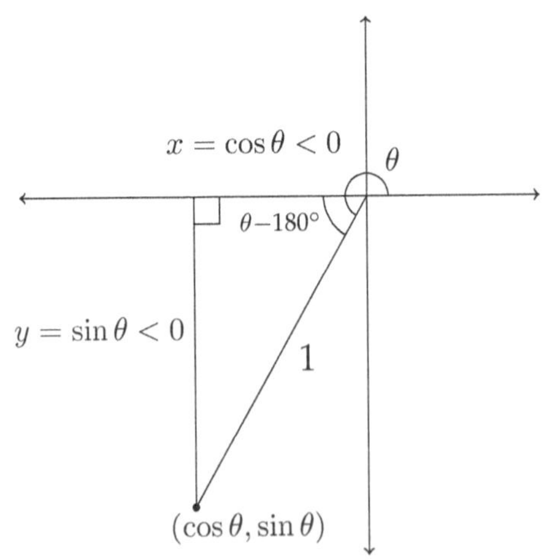

In Quadrant III, the cosine and sine are both negative.

Quadrant IV

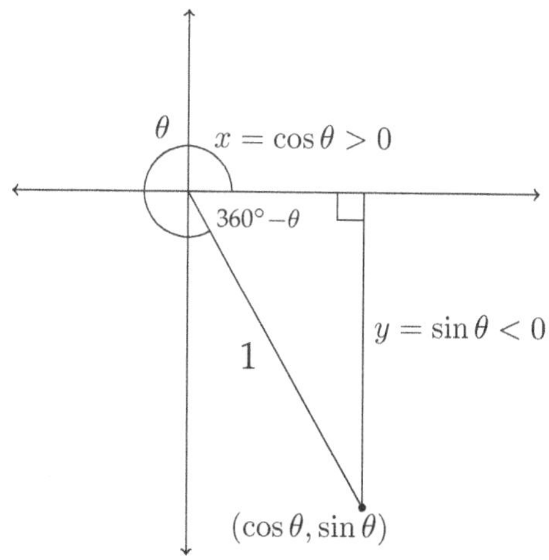

In Quadrant IV, the cosine is positive and the sine is negative.

The process for finding reference angles depends on which quadrant the angle terminates in.

Examples

Find the reference angle for the following angles:

1. $128°$ 2. $241°$ 3. $327°$

1. An angle of 128°terminates in Quadrant II. To find the reference angle, we would subtract the angle from 180°: $180° - 128° = 52°$.

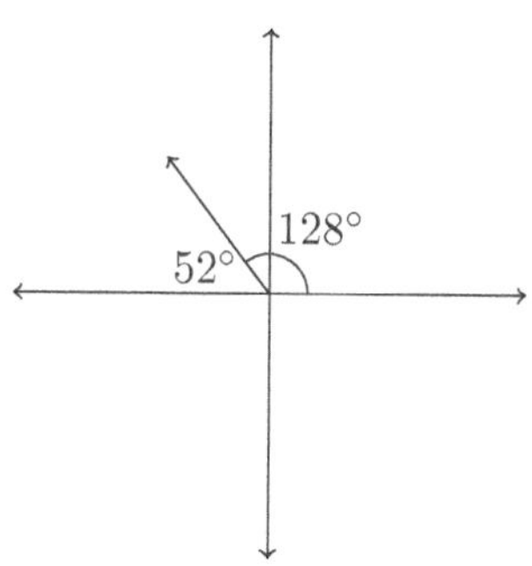

2. An angle of 241°terminates in Quadrant III. To find the reference angle, we would subtract 180°from the angle: $241^\circ - 180^\circ = 61^\circ$.

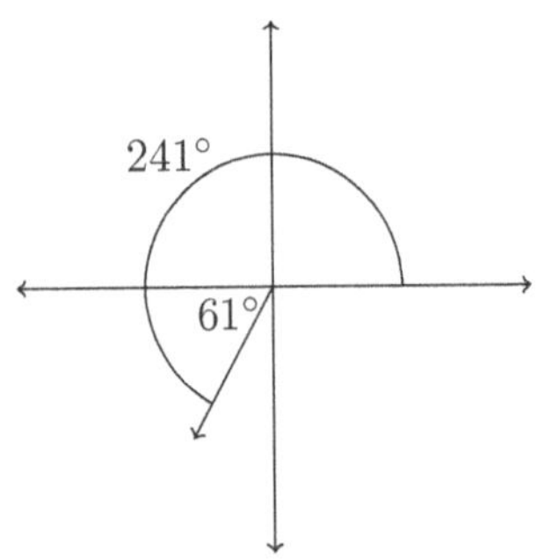

3. An angle of 327°terminates in Quadrant IV. To find the reference angle, we subtract the angle from 360°: $360^\circ - 327^\circ = 33^\circ$.

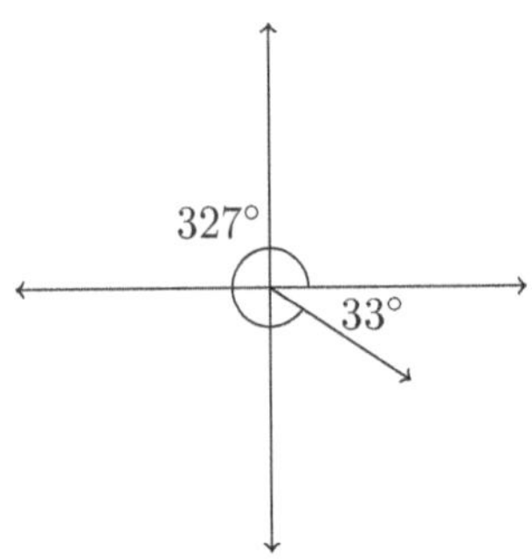

Once we know the reference angle, we can find the trigonometric functions for the original angle itself. In example 1, we had 128°, an angle in Quadrant II with a reference angle of 52°. Therefore, if we want to find the sine, cosine and tangent of 128°, then we should find the sine, cosine and tangent of 52°and apply the appropriate positive or negative sign.

Example 1 Quadrant II

In Quadrant II, x-coordinates are negative and y-coordinates are positive. This means that $\cos\theta < 0$ and $\sin\theta > 0$. The values for this process are given below:

$\sin 52^\circ \approx 0.7880$	$\sin 128^\circ \approx 0.7880$
$\cos 52^\circ \approx 0.6157$	$\cos 128^\circ \approx -0.6157$
$\tan 52^\circ \approx 1.280$	$\tan 128^\circ \approx -1.280$

Example 2 Quadrant III

In Quadrant III, x-coordinates are negative and y-coordinates are also negative. This means that $\cos\theta < 0$ and $\sin\theta < 0$. The values for this process are given below:

$\sin 61^\circ \approx 0.8746$	$\sin 241^\circ \approx -0.8746$
$\cos 61^\circ \approx 0.4848$	$\cos 241^\circ \approx -0.4848$
$\tan 61^\circ \approx 1.8040$	$\tan 241^\circ \approx 1.8040$

Example 3 Quadrant IV

In Quadrant IV, x-coordinates are positive and y-coordinates are negative. This means that $\cos\theta > 0$ and $\sin\theta < 0$. The values for this process are given below:

$\sin 33^\circ \approx 0.5446$	$\sin 327^\circ \approx -0.5446$
$\cos 33^\circ \approx 0.8387$	$\cos 327^\circ \approx 0.8387$
$\tan 33^\circ \approx 0.6494$	$\tan 327^\circ \approx -0.6494$

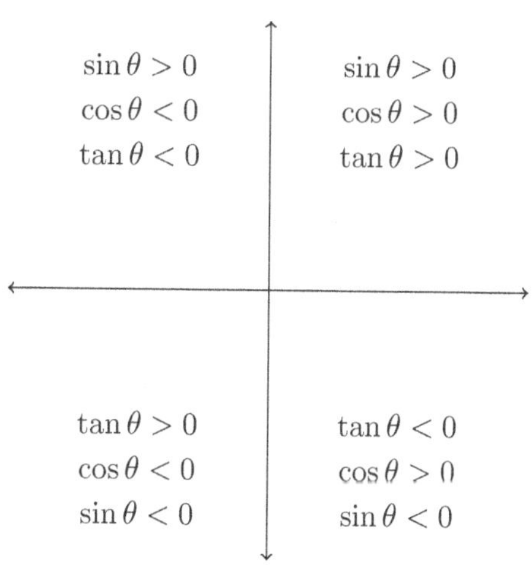

In Quadrant I, ALL the trigonometric functions are positive.

In Quadrant II, the SIN function is positive (as well as the CSC).

In Quadrant III, the TAN function is positive (as well as the COT).

In Quadrant IV, the COS function is positive (as well as the SEC).

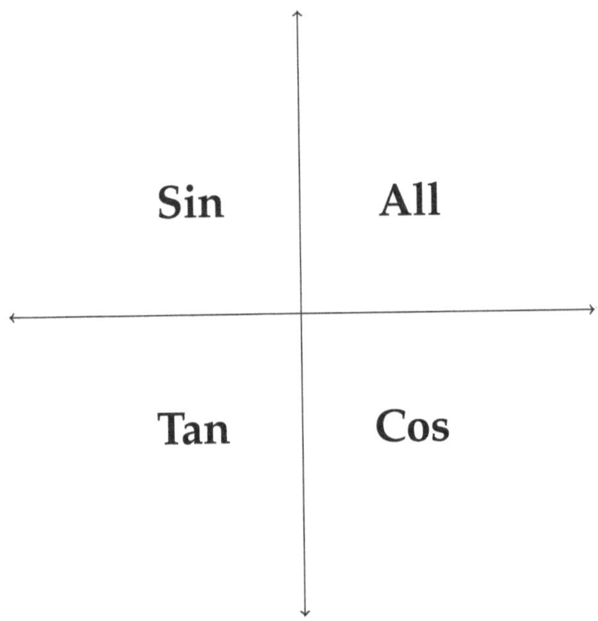

A common mneumonic device to remember these relationships is the phrase: "**A**ll **S**tudents **T**ake **C**alculus." This can help you remember which trigonometric functions are positive in each of the four quadrants.

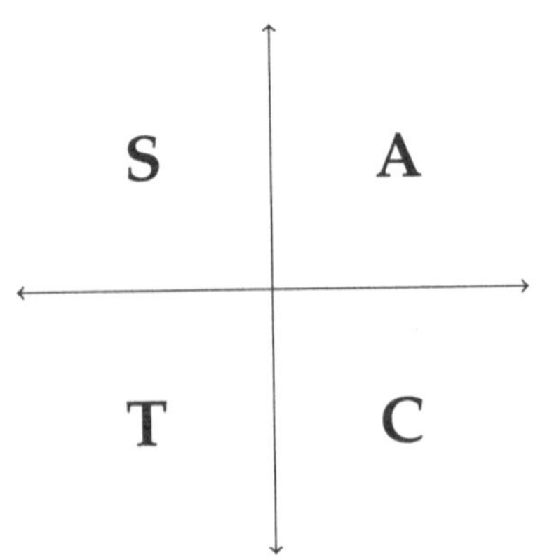

Reference Angles for Negative Angles

Negatively measured angles rotate in a clockwise direction.

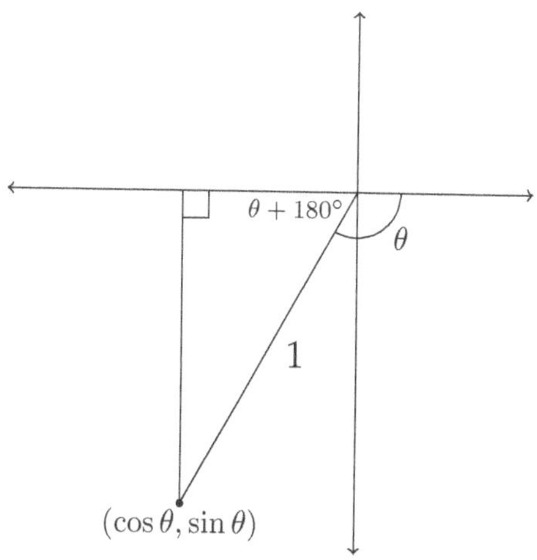

There are a variety of methods for finding the reference angle for a negatively valued angle. You can find a positive angle that is co-terminal with the negative angle and then find the reference angle for the positive angle. You can also drop a perpendicular to the x-axis to find the reference angle for the negative angle directly.

For example, the angle $-120°$ terminates in Quadrant III and is co-terminal with the positive angle 240°. Either way, when you drop a perpendicular to the x-axis, you find that the reference angle is 60°.

If you are given the value of one of the trigonometric functions of an angle θ, and know which quadrant θ is located in, you can find the other trigonometric functions for that angle.

Example

Given θ in Quadrant IV with $\cos\theta = \frac{1}{5}$, find $\sin\theta$ and $\tan\theta$.

If $\cos\theta = \frac{1}{5}$, then the adjacent side and the hypotenuse must be in a ratio of 1:5.

We can label these sides as 1 and 5 and then find the length of the third side in the triangle. This will allow us to find $\sin\theta$ and $\tan\theta$.

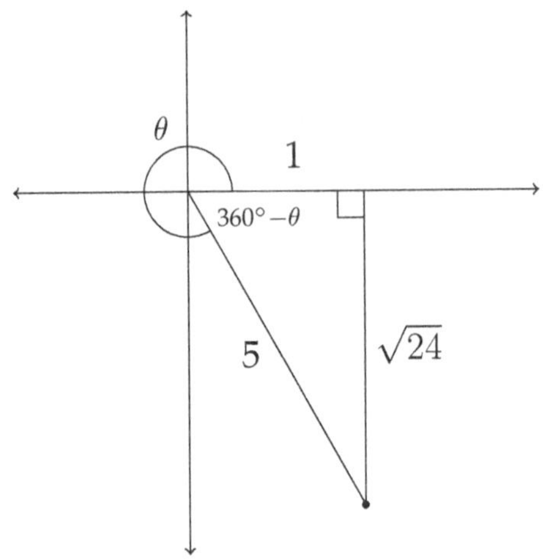

Using the Pythagorean Theorem:

$$5^2 = 1^2 + s^2$$

$$25 - 1 = s^2$$

$$\sqrt{24} = s$$

we find that the side opposite the reference angle for θ is $\sqrt{24}$ or $2\sqrt{6}$. We can now find $\sin\theta$ and $\tan\theta$:

$$\sin\theta = \frac{\sqrt{24}}{5}$$

and

$$\tan\theta = \frac{\sqrt{24}}{1} = \sqrt{24}$$

In the problems in this section, the reciprocal functions secant, cosecant and cotangent are used. Remember that:

$$\sec\theta = \frac{1}{\cos\theta} = \frac{hyp}{adj}$$

$$\csc\theta = \frac{1}{\sin\theta} = \frac{hyp}{opp}$$

$$\cot\theta = \frac{1}{\tan\theta} = \frac{adj}{opp}$$

Exercises 2.1

Determine the quadrant in which the angle θ lies.

1. $\cos\theta > 0, \tan\theta > 0$
2. $\sin\theta < 0, \cos\theta > 0$
3. $\sec\theta > 0, \tan\theta < 0$
4. $\cot\theta > 0, \cos\theta < 0$
5. $\sin\theta > 0, \cos\theta < 0$
6. $\sin\theta > 0, \cot\theta > 0$
7. $\sin\theta < 0, \cos\theta < 0$
8. $\csc\theta > 0, \cot\theta < 0$

Determine which quadrant the given angle terminates in and find the reference angle for each.

9. $195°$
10. $330°$
11. $120°$
12. $210°$
13. $135°$
14. $300°$
15. $-100°$
16. $225°$
17. $315°$
18. $\frac{5\pi}{4}$
19. $-\frac{2\pi}{3}$
20. $\frac{7\pi}{3}$
21. $\frac{11\pi}{4}$
22. $\frac{7\pi}{6}$
23. $\frac{11\pi}{6}$

Find $\sin\theta$, $\cos\theta$ and $\tan\theta$ in each problem.

24. $\sin\theta = -\frac{12}{13}$, θ in Quadrant IV

25. $\cos\theta = -\frac{4}{5}$, θ in Quadrant II

26. $\cos\theta = \frac{1}{4}$, θ in Quadrant I

27. $\tan\theta = \frac{3}{2}$, θ in Quadrant III

28. $\tan\theta = -\frac{4}{5}$, θ in Quadrant II

29. $\sin\theta = \frac{3}{8}$, θ in Quadrant II

30. $\sin\theta = -\frac{1}{3}$, θ in Quadrant III

31. $\tan\theta = 5$, θ in Quadrant I

32. $\sec\theta = -2, \tan\theta < 0$

33. $\cot\theta = \sqrt{3}, \cos\theta < 0$

34. $\tan\theta = -\frac{1}{3}, \sin\theta < 0$

35. $\csc\theta = \sqrt{2}, \cos\theta > 0$

36. $\cos\theta = -\frac{2}{5}, \tan\theta > 0$

37. $\sec\theta = 2, \sin\theta < 0$

38. $\sin\theta = \frac{1}{\sqrt{2}}, \cos\theta > 0$

39. $\sin\theta = -\frac{2}{3}, \cot\theta > 0$

2.2 Graphing Trigonometric Functions

We have seen how to determine the values of trigonometric functions for angles terminating in Quadrants II, III, and IV. This allows us to make a graph of the values of the sine function for any angle. In the chart below, I have listed the values for the sine function for angles between 0°and 360°.

θ	$\sin\theta$		θ	$\sin\theta$
0°	$=0$		100°	≈ 0.9848
10°	≈ 0.1737		110°	≈ 0.9397
20°	≈ 0.3420		120°	≈ 0.8660
30°	$=0.5$		130°	≈ 0.7660
40°	≈ 0.6428		140°	≈ 0.6428
50°	≈ 0.7660		150°	$=0.5$
60°	≈ 0.8660		160°	≈ 0.3420
70°	≈ 0.9397		170°	≈ 0.1737
80°	≈ 0.9848		180°	$=0$
90°	$=1$			

θ	$\sin\theta$		θ	$\sin\theta$
180°	$=0$		280°	≈ -0.9848
190°	≈ -0.1737		290°	≈ -0.9397
200°	≈ -0.3420		300°	≈ -0.8660
210°	$=-0.5$		310°	≈ -0.7660
220°	≈ -0.6428		320°	≈ -0.6428
230°	≈ -0.7660		330°	$=-0.5$
240°	≈ -0.8660		340°	≈ -0.3420
250°	≈ -0.9397		350°	≈ -0.1737
260°	≈ -0.9848		360°	$=0$
270°	$=-1$			

On the next page we see a graph of these points plotted on the coordinate axes.

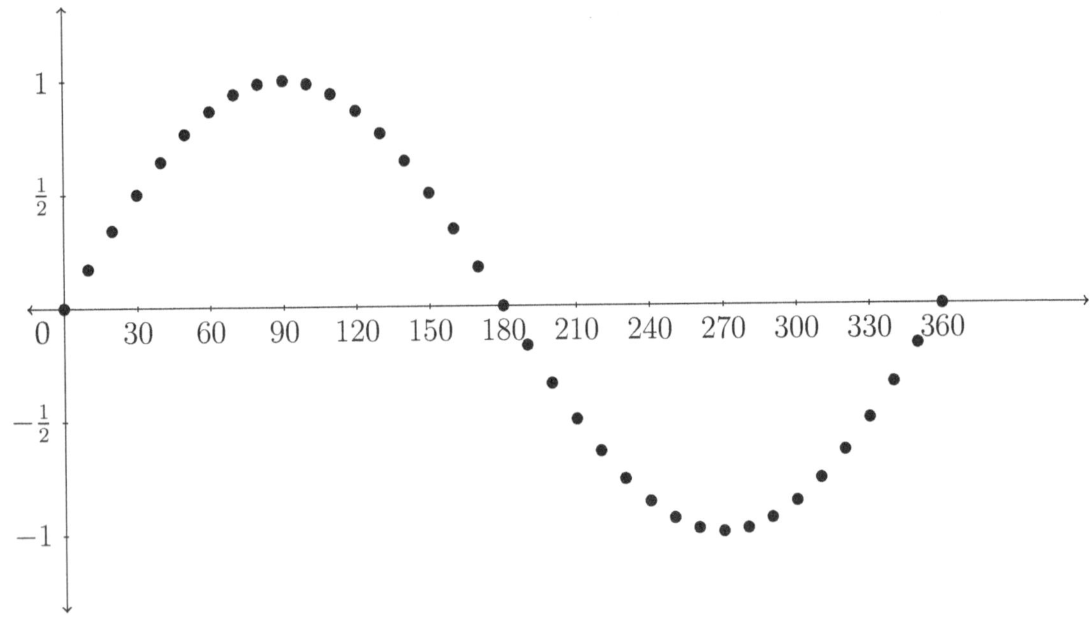

In graphing trigonometric functions, we typically use radian measure along the x-axis, so the graph would generally look like this:

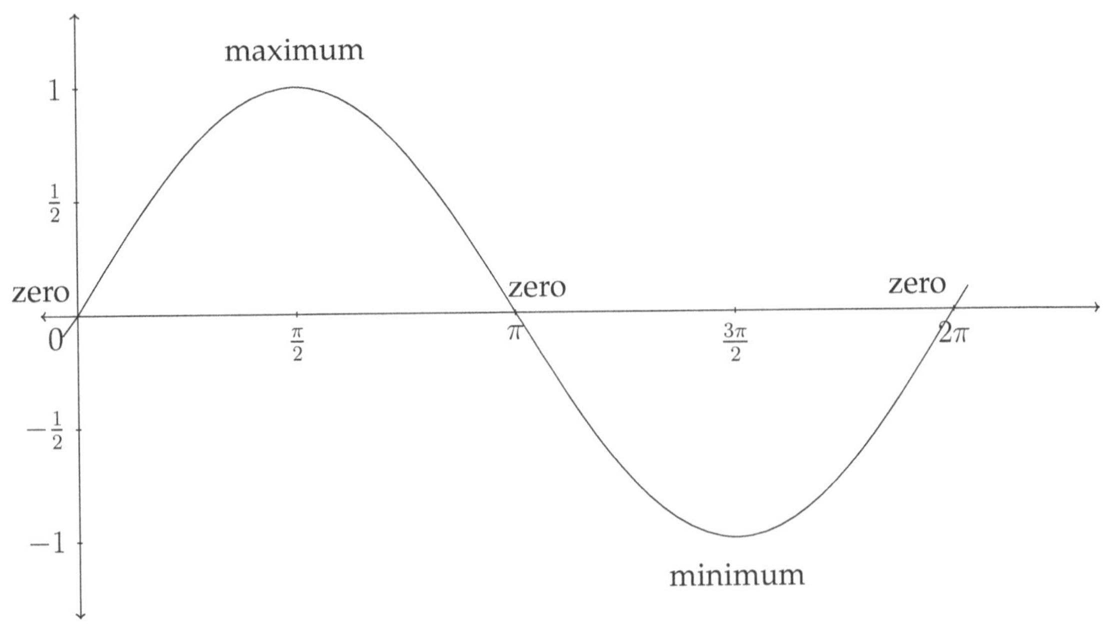

The graph of the standard sine function begins at the zero point, then rises to the maximum value of 1 between 0 and $\frac{\pi}{2}$ radians. It then decreases back to 0 at π radians before crossing over into the negative values and hitting its minimum value at $\frac{3\pi}{2}$ radians. It then goes back up to 0 at 2π radians before starting all over again.

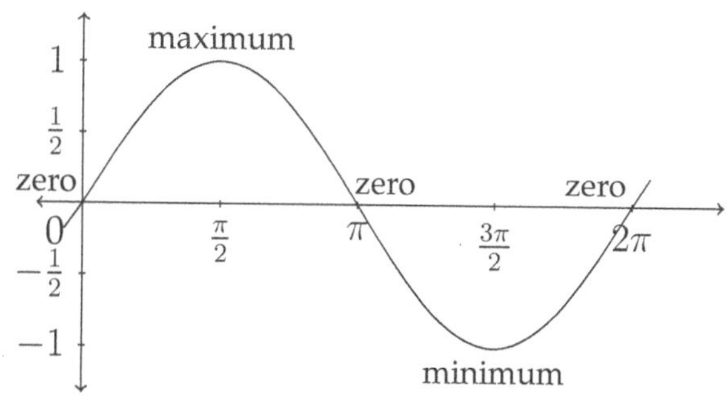

The standard cosine graph behaves in a similar but slightly different way. We saw earlier that $\cos 0° = 1$, so the cosine graph would start at the point $(0, 1)$, then gradually decrease to zero. A picture of the standard cosine graph would look like the figure below:

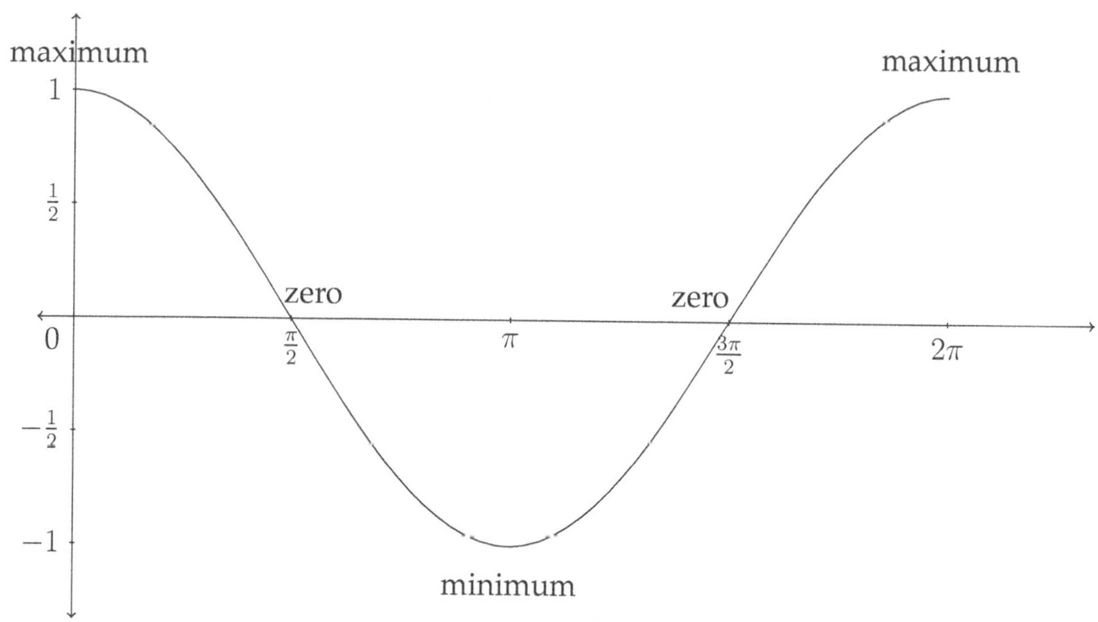

The sine and cosine graphs are sometimes referred to as a "sine wave" or "sinusoid" and can be very useful in modeling phenomena that occur in waves. Examples of this are the rise and fall of the tides; sound waves and music; electricity; and the length of day throughout the year. The standard sine and cosine graphs must be modified to fit a particular application so that they will effectively model the situation. The ideas that we examine next will explain how to modify the sine and cosine graphs to fit a variety of different situations.

There are four aspects to the sine and cosine functions to take into consideration when making a graph. These are:

1) The Amplitude of the graph
2) The Period of the graph
3) The Vertical Shift of the graph
4) The Phase Shift of the graph

Amplitude

The amplitude of a sine or cosine function refers to the maximum and minimum values of the function. In the standard sine and cosine graphs, the maximum value is 1 and the minimum value is -1. The amplitude is one-half the difference between the maximum and minimum values. In the standard graphs the difference between the maximum and minimum is $1-(-1)=2$; one-half of this is 1, so the amplitude of the standard sine and cosine functions is 1.

The value of the amplitude is also the absolute value of the coefficient of the sine or cosine expression. In the standard graph, $y=\sin x$, the coefficient of the sine function is 1, so the amplitude is 1. In the function $y=2\sin x$, all the y values will be multiplied by 2 and the amplitude of the function will be 2. The graph for $y=2\sin x$ is shown on the next page.

$$y = 2\sin x$$

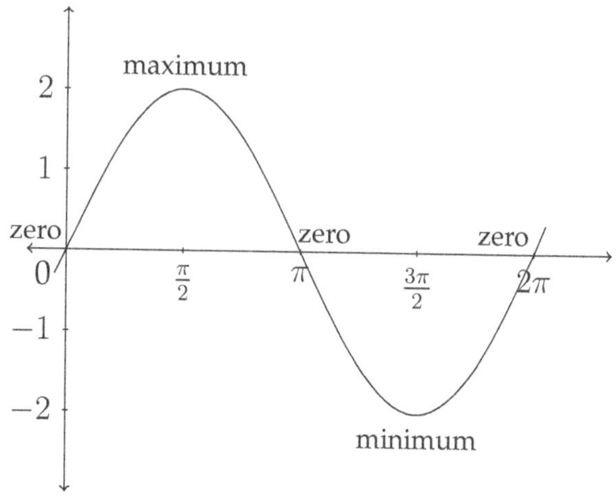

A negative value of the coefficient in front of a trigonometric function will not change the Amplitude of the function, but it will change the shape of the function. For example, the function:

$$y = -\sin x$$

has an amplitude of 1, but the graph will be different from the graph $y = \sin x$. All of the y-values of the function $y = -\sin x$ will have the opposite sign as the y-values of the function $y = \sin x$. The graph for $y = -\sin x$ appears below:

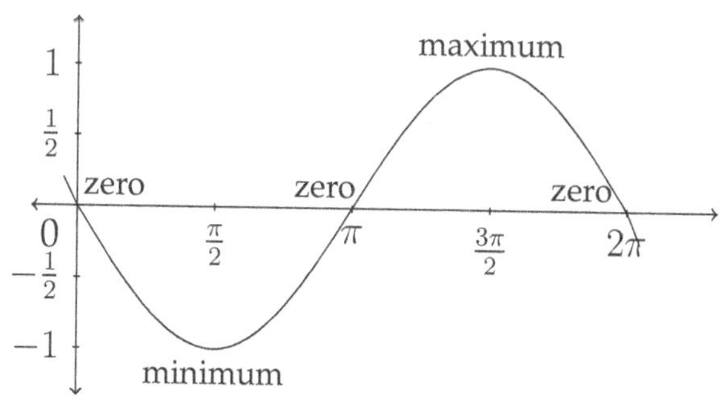

Notice that, because of the negation of the y-values, the graph begins at 0, as does the standard sine function, but the graph of $y = -\sin x$ first goes to a minimum value before crossing through 0 again up to the maximum value.

Likewise, the graph of $y = -\cos x$ begins at the minimum value before crossing through 0 and going to the maximum value, back through 0 and ending at the minimum value again.

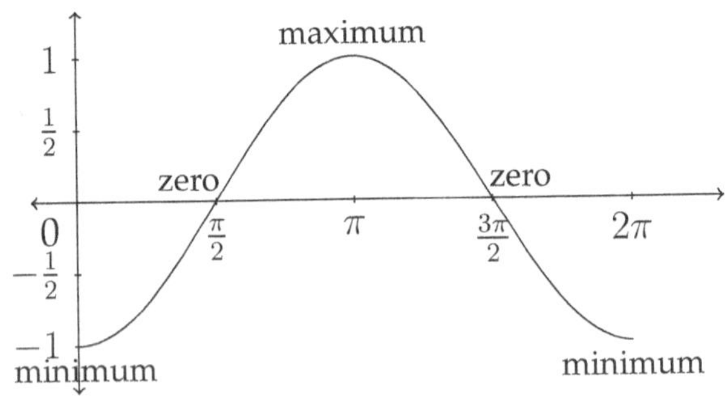

$y = -\cos x$

Period

The period of the graph refers to how long it takes the graph to complete one full cycle of values. In the standard sine and cosine functions, the period is 2π radians. The function completes a single "wave" and returns to its starting place between 0 and 2π. A coefficient in front of the variable in a sine or cosine function will affect the period of the graph. In the general expression $y = A \sin Bx$, the value of A affects the amplitude of the function and the value of B affects the period of the function.

If we examine the table of values for the standard sine function, we can see how the coefficient of the x-variable will affect the period of the graph. Starting with the table from the standard sine function:

θ	$\sin\theta$		θ	$\sin\theta$
0°	$= 0$		100°	≈ 0.9848
10°	≈ 0.1737		110°	≈ 0.9397
20°	≈ 0.3420		120°	≈ 0.8660
30°	≈ 0.5		130°	≈ 0.7660
40°	≈ 0.6428		140°	≈ 0.6428
50°	≈ 0.7660		150°	≈ 0.5
60°	≈ 0.8660		160°	≈ 0.3420
70°	≈ 0.9397		170°	≈ 0.1737
80°	≈ 0.9848		180°	$= 0$
90°	$= 1$			

θ	$\sin\theta$		θ	$\sin\theta$
180°	$= 0$		280°	≈ -0.9848
190°	≈ -0.1737		290°	≈ -0.9397
200°	≈ -0.3420		300°	≈ -0.8660
210°	≈ -0.5		310°	≈ -0.7660
220°	≈ -0.6428		320°	≈ -0.6428
230°	≈ -0.7660		330°	$= -0.5$
240°	≈ -0.8660		340°	≈ -0.3420
250°	≈ -0.9397		350°	≈ -0.1737
260°	≈ -0.9848		360°	$= 0$
270°	$= -1$			

If we create a similar table for the function $y = \sin(2x)$, then we can see how this will affect the graph:

θ	2θ	$\sin(2\theta)$		θ	2θ	$\sin(2\theta)$
0°	0°	$= 0$		100°	200°	≈ -0.3420
10°	20°	≈ 0.3420		110°	220°	≈ -0.6428
20°	40°	≈ 0.6428		120°	240°	≈ -0.8660
30°	60°	≈ 0.8660		130°	260°	≈ -0.9848
40°	80°	≈ 0.9848		140°	280°	≈ -0.9848
50°	100°	≈ 0.9848		150°	300°	≈ -0.8660
60°	120°	≈ 0.8660		160°	320°	≈ -0.6428
70°	140°	≈ 0.6428		170°	340°	≈ 0.3420
80°	160°	≈ 0.3420		180°	360°	$= 0$
90°	180°	$= 0$				

In the previous table we can see that the function $y = \sin(2x)$ completes one full cycle between 0 and π radians instead of the the standard 0 to 2π radians. The graph for these points is shown below. The coordinates for the x-values between π and 2π radians are shown as well.

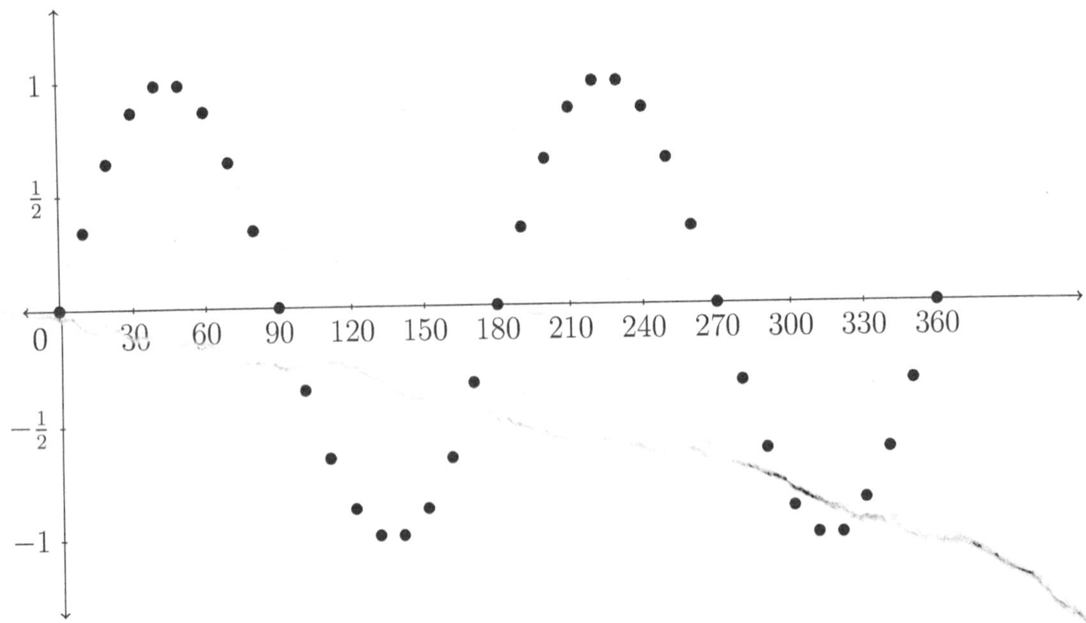

In this graph, you can see that there are two complete waves between 0 and 2π radians, or one complete wave between 0 and π radians. So, in a sine or cosine function of the form $y = A \sin Bx$, the amplitude will be $|A|$ and the period will be $\frac{2\pi}{B}$. The standard graph for one complete cycle of the function $y = \sin(2x)$ is shown below:

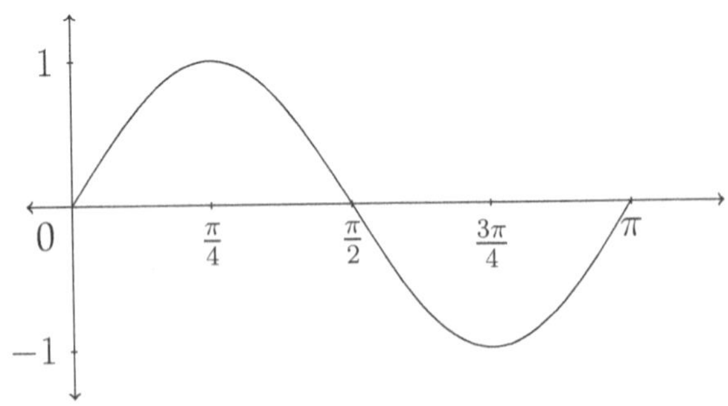

Notice that, because the period has been cut in half, the x-coordinates that correspond to the maximum, minimum, and zero y-coordinates are cut in half as well. Let's look at some examples of how the Amplitude and the Period affect the graphs of the sine and cosine functions.

Example 1

Graph one full period of the function $y = -2\sin 3x$.

The amplitude in this case is 2, but since the coefficient is negative, this sine graph will begin by first going to the minimum value. The period of the graph will be $\frac{2\pi}{B}$, or in this case $\frac{2\pi}{3}$ instead of 2π. To determine the x-values for the maximum, minimum and zero y-values, we should examine how these are determined for the standard sine curve.

The maximum, minimum and zero y-values for a standard sine curve occur at the quadrantal angles, that is to say, the angles that separate the four quadrants from each other. The quadrantal angles are 0°or 0 radians, 90°or $\frac{\pi}{2}$ radians, 180°or π radians, 270°or $\frac{3\pi}{2}$ radians and 360°or 2π radians. These x-values produce the "critical" y-values of the zero, maximum and minimum.

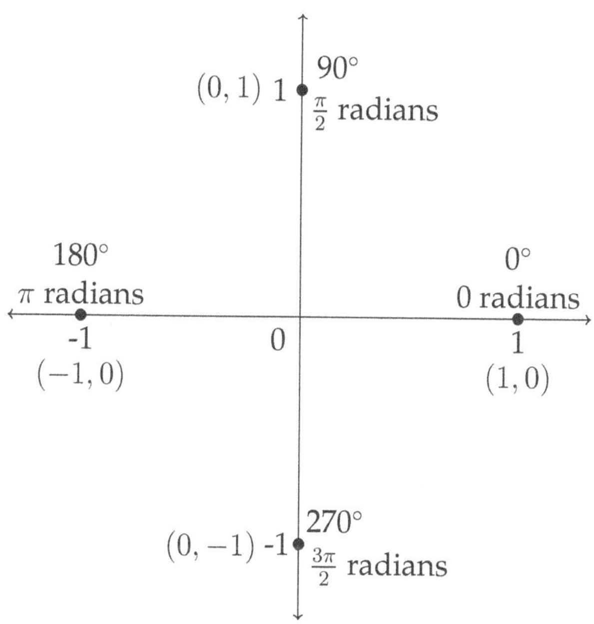

In the standard sine or cosine graph, the distance from each "critical value" of the graph to the next is always a "jump" of $\frac{\pi}{2}$ along the x-axis. This is one-fourth of the period: $\frac{2\pi}{1} * \frac{1}{4} = \frac{\pi}{2}$. So, to determine the labels for the critical values of the graph along the x-axis, we should take the new period and multiply by $\frac{1}{4}$.

The function we are working with is $y = -2\sin 3x$, so to find the new period we calculated $\frac{2\pi}{B}$, which was $\frac{2\pi}{3}$. Then, in order to label the x-axis properly we should next take $\frac{2\pi}{3}$ and multiply by $\frac{1}{4}$.

$$\frac{2\pi}{3} * \frac{1}{4} = \frac{2\pi}{12} = \frac{\pi}{6}$$

So, the critical values along the x-axis will be:

$$\frac{1\pi}{6}, \frac{2\pi}{6}, \frac{3\pi}{6}, \text{ and } \frac{4\pi}{6}$$

We want to express these in lowest terms, so we would label them as $\frac{\pi}{6}, \frac{\pi}{3}, \frac{\pi}{2}$, and $\frac{2\pi}{3}$. The graph will start at zero, then (because the value of the coefficient A is negative) it will go down to a minimum value at $\frac{\pi}{6}$, back to zero at $\frac{\pi}{3}$, then up to the maximum at $\frac{\pi}{2}$ and back down to zero at $\frac{2\pi}{3}$ to complete one full period of the graph. The graph for this function is pictured below. Notice that the minimum y-value is -2 and the maximum y-value is 2 because $A = 2$.

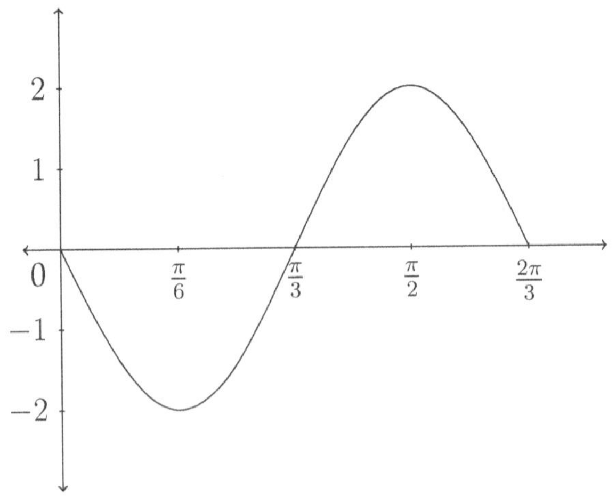

$y = -2\sin 3x$

Let's look at an example using the cosine graph.

Example 2

Graph one full period of the function $y = 5\cos\frac{2}{3}x$.

The amplitude of the function is 5 because $A = 5$, so the maximum y-value will be 5 and the minimum y-value will be -5. The period of the graph will be $\frac{2\pi}{B}$, which in this case is $\frac{2\pi}{\frac{2}{3}} = 2\pi * \frac{3}{2} = 3\pi$. So the period is 3π. The critical values along the x-axis will start at 0 and be separated by "jumps" of $3\pi * \frac{1}{4} = \frac{3\pi}{4}$. So the critical values along the x-axis will be:

$$0, \frac{3\pi}{4}, \frac{6\pi}{4}, \frac{9\pi}{4}, \text{ and } \frac{12\pi}{4}$$

We want to express these in lowest terms so we would label them as $\frac{3\pi}{4}$, $\frac{3\pi}{2}$, $\frac{9\pi}{4}$, and 3π. The graph will start at the maximum y-value of 5 at $x = 0$, then it will go to zero at $x = \frac{3\pi}{4}$, down to the minimum y-value of -5 at $x = \frac{3\pi}{2}$, back through 0 at $x = \frac{9\pi}{4}$, and then up to the maximum y-value of 5 at $x = 3\pi$ to complete one full period of the graph. The graph of $y = 5\cos\frac{2}{3}x$. is shown below.

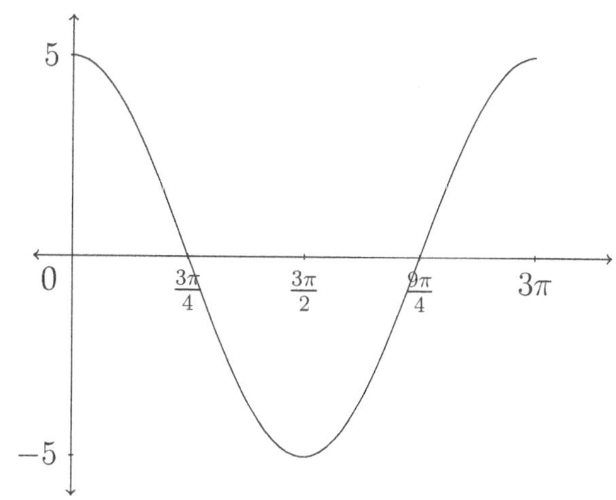

$y = 5\cos\frac{2}{3}x$

Determining an equation from a graph

Sometimes, you will be given a graph and asked to determine an equation which satisfies the conditions visible in the graph. So far, we have only discussed two of the possible transformations of a trigonometric function - the amplitude and period. Remember that in an equation of the form $y = A\sin Bx$ or $y = A\cos Bx$, the amplitude is $|A|$ and the period is $\frac{2\pi}{B}$. So, to write an equation for a trigonometric function, we need to determine the values of A and B.

Example 3

Deterimine an equation that satisfies the given graph.

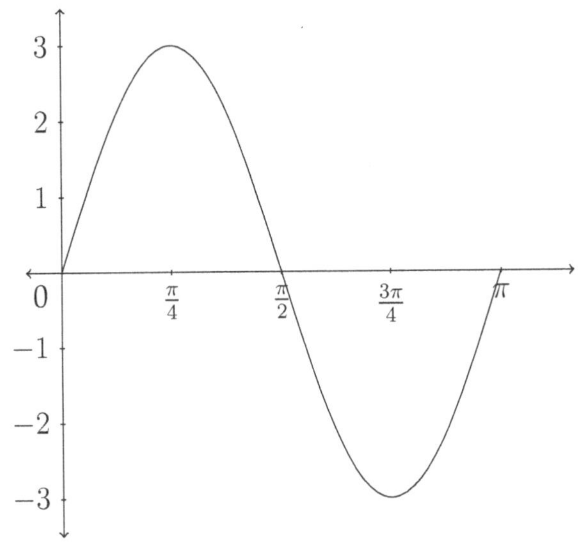

First note that the maximum y-value for the graph is 3 and the minimum is -3. This means that the amplitude is 3. Next we see that there is one complete period of the function between 0 and π, this means that the period is π. From this information, we know that $A = 3$ and that the period for the graph is π. Since the period $P = \frac{2\pi}{B}$, then we know that $B = \frac{2\pi}{P}$. So, $B = \frac{2\pi}{\pi} = 2$.

Lastly, we notice that this graph starts with a y-value of 0, then goes to the maximum, back through 0 to the minimum and then back to zero to form one complete wave. This is the signature of a sine function, so the answer to this problem would be:

$$y = 3\sin 2x$$

Example 4

Deterimine an equation that satisfies the given graph.

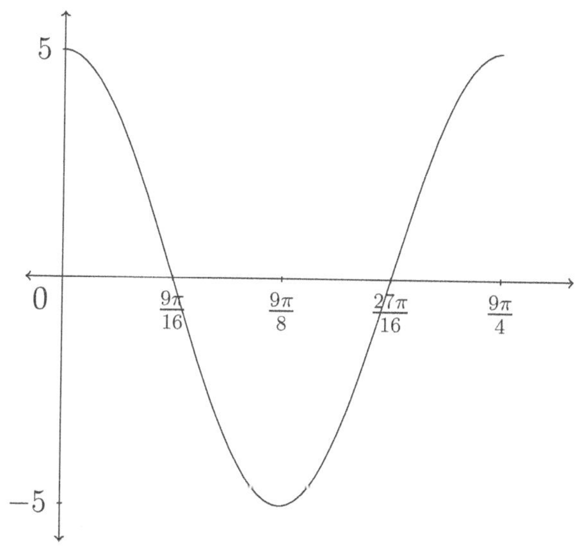

First note that the maximum y-value for the graph is 5 and the minimum is -5. This means that the amplitude is 5. Next we see that there is one complete period of the function between 0 and $\frac{9\pi}{4}$, this means that the period is $\frac{9\pi}{4}$. From this information, we know that $A = 5$ and that the period for the graph is $\frac{9\pi}{4}$. Since $P = \frac{2\pi}{B}$, and $B = \frac{2\pi}{P}$, then $B = \frac{2\pi}{(\frac{9\pi}{4})} = 2\pi * \frac{4}{9\pi} = \frac{8}{9}$.

Lastly, this graph starts with a y-value of 5, which is the maximum y-value. It then goes to 0 and down to the minimum, back through 0 and then back to the maximum to form one complete wave. Since this is the signature of the cosine function, the answer to this problem would be:

$$y = 5\cos\frac{8}{9}x$$

Exercises 2.2

Determine the amplitude and the period for each problem and graph one period of the function. Identify important points on the x and y axes.

1. $y = \cos 4x$
2. $y = -\sin 2x$
3. $y = 3\sin 3x$
4. $y = -2\cos 5x$
5. $y = 4\cos \frac{1}{2}x$
6. $y = 2\sin \frac{1}{3}x$
7. $y = -\frac{1}{2}\sin \frac{2}{3}x$
8. $y = -3\cos \frac{3}{5}x$
9. $y = -4\sin 6x$
10. $y = 3\sin 4x$
11. $y = 2\cos \frac{3}{2}x$
12. $y = 3\cos \frac{5}{3}x$

Determine an equation that satifies the given graph.

13.

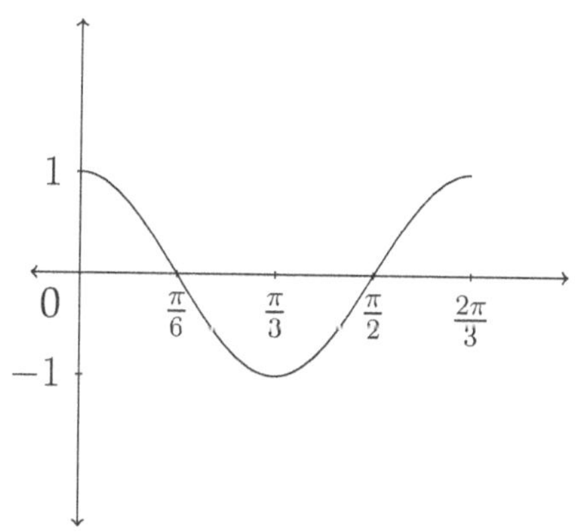

14.

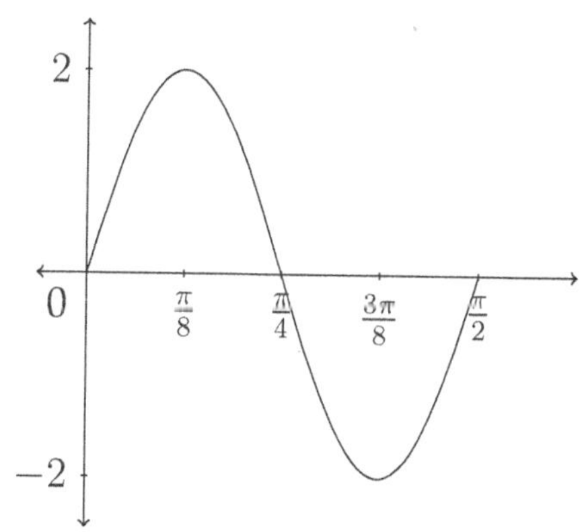

15. 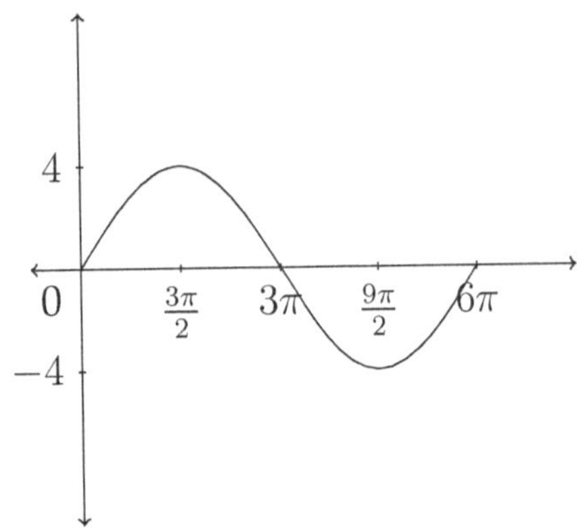

16. 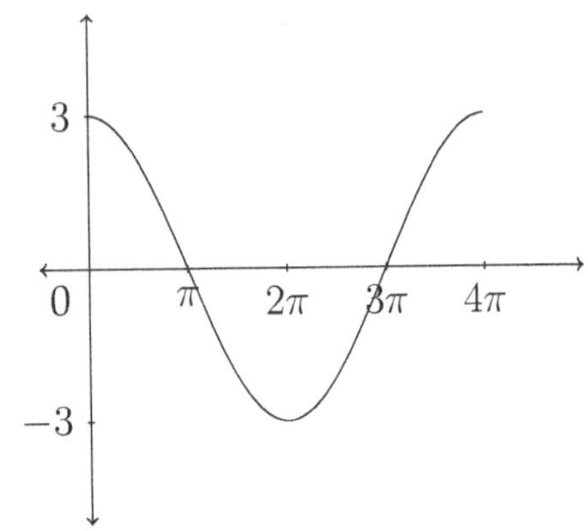

17. 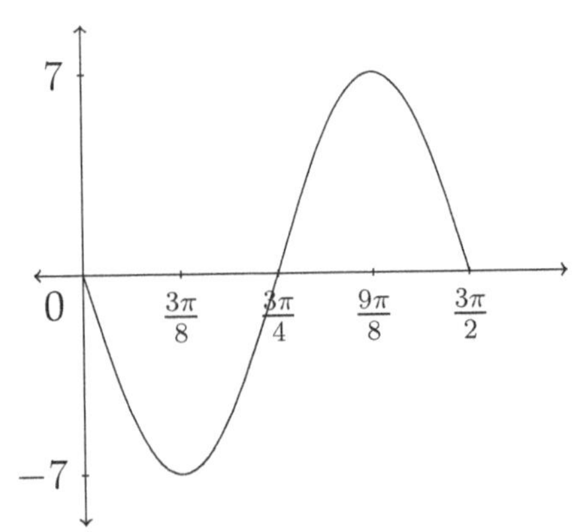

18. 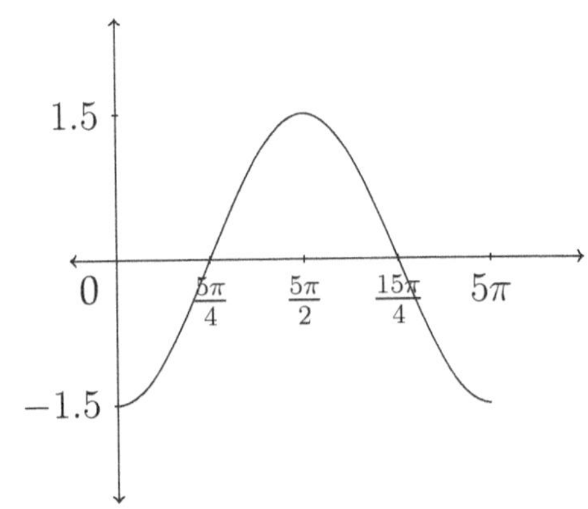

19. 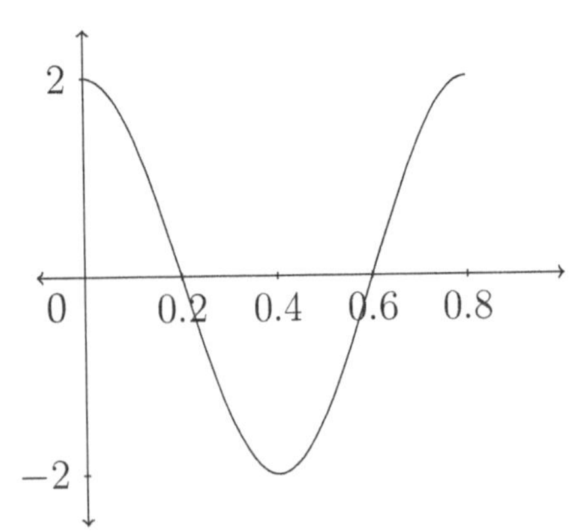

20. 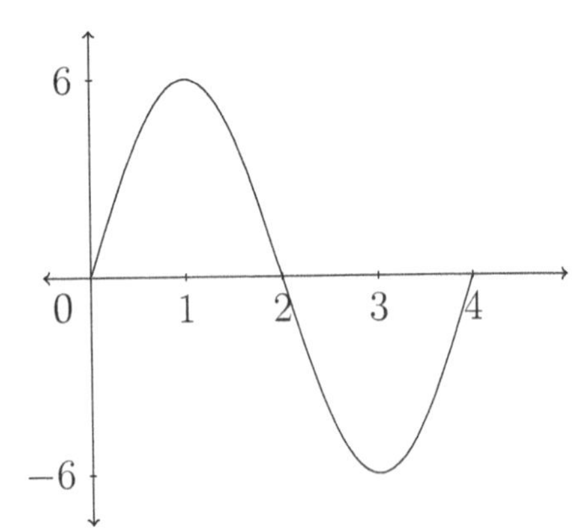

2.3 The Vertical Shift of a Trigonometric Function

If a constant is added or subtracted to a trigonometric function, this will affect the y-values of the function. If we consider the function $y = 5 + \sin x$, then each of the standard y-values would have 5 added to it, which would shift the graph up 5 units.

The chart below considers just the quadrantal values for the sine function:

θ	$\sin\theta$	$5+\sin\theta$
0	0	5
$\pi/2$	1	6
π	0	5
$3\pi/2$	-1	4
2π	0	5

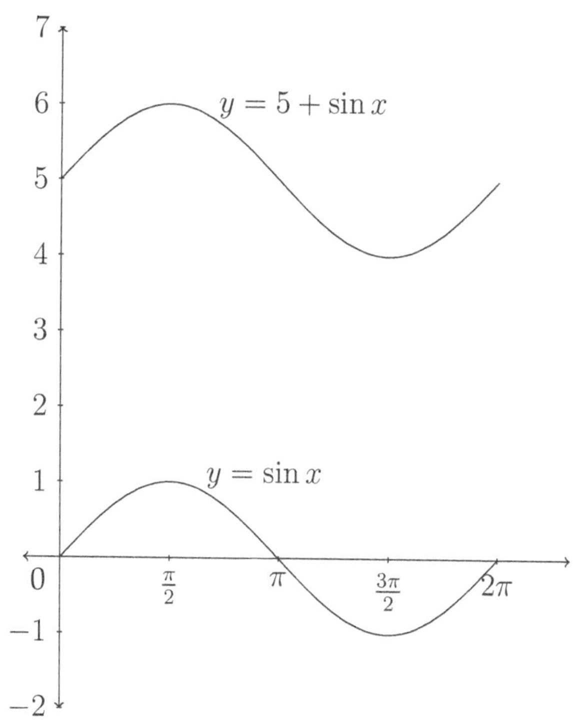

Sometimes the x-axis is drawn through the line that is the new "zero" or "mid-line" for the function - in this case it would be $y = 5$.

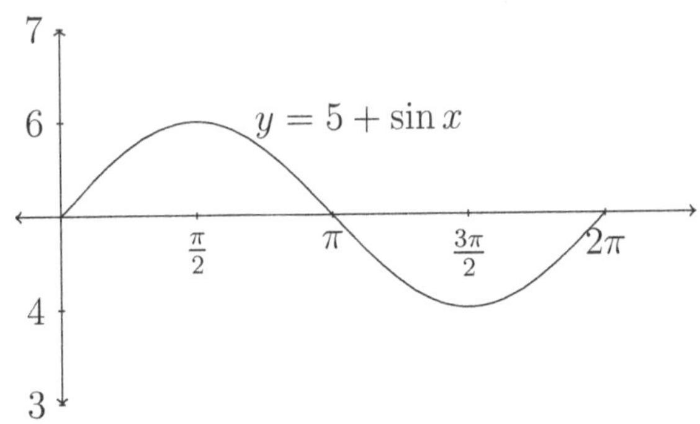

Likewise, a negative constant would move the graph down, as each y-value would be less than the corresponding y-value in the standard sine function.

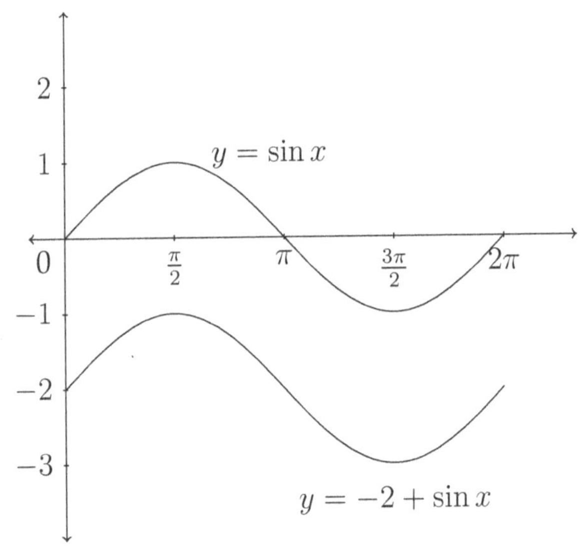

In the previous examples the constant has been written in front of the sine function for clarity. Often the constant is written after the function:

$$y = \sin x + 5$$

or

$$y = \sin x - 2$$

We have now examined three of the four transformations of trigonometric functions that are discussed in this chapter - amplitude, period and vertical shift. A general equation for a sinusoid that involves these three transformations would be:

$$y = A\sin(Bx) + D$$
$$\text{or}$$
$$y = A\cos(Bx) + D$$

In determining an equation from a graph that involves a vertical shift, the value of A will be half the distance between the maximum and minimum values:

$$A = \frac{max - min}{2}$$

and the value of D will be the average of the maximum and minimum values:

$$D = \frac{max + min}{2}$$

Example

Determine an equation that satisfies the given graph.

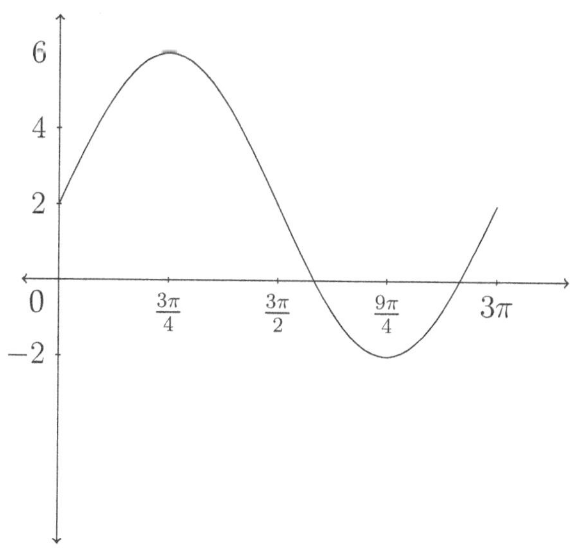

In this graph, the maximum y-value is 6 and the minimum y-value is -2. The average of these two:

$$\frac{max+min}{2}=\frac{6+(-2)}{2}=\frac{4}{2}=2=D$$

is the value of D, the vertical shift.

The distance between 6 and -2 is $6-(-2)=8$. Half the distance between the max and min is 4, which is the value of A.

$$\frac{max-min}{2}=\frac{6-(-2)}{2}=\frac{8}{2}=4=A$$

The graph completes one full cycle between 0 and 3π, so the period would be 3π and the value of B would be $B=\frac{2\pi}{P}=\frac{2\pi}{3\pi}=\frac{2}{3}=B$. So a correct equation for the graph would be:

$$y=4\sin\frac{2}{3}x+2$$

Exercises 2.3

Determine the Amplitude, Period and Vertical Shift for each function below and graph one period of the function. Identify the important points on the x and y axes.

1. $y = \sin x + 1$
2. $y = \cos x - 1$
3. $y = 2\cos x - \frac{1}{2}$
4. $y = 5\sin x + 4$
5. $y = -\sin(\frac{1}{4}x) + 1$
6. $y = -\cos(2x) + 7$
7. $y = \frac{1}{3}\sin(\pi x) - 4$
8. $y = -\frac{1}{2}\cos(2\pi x) + 2$
9. $y = 5\cos(\frac{1}{2}x) + 1$
10. $y = 4\sin(\frac{1}{3}x) - 1$
11. $y = 3\cos x + 2$
12. $y = 2\sin x + 3$
13. $y = 2 - 4\cos(3x)$
14. $y = 5 - 3\sin(2x)$

Determine an equation that satifies the given graph.

15.

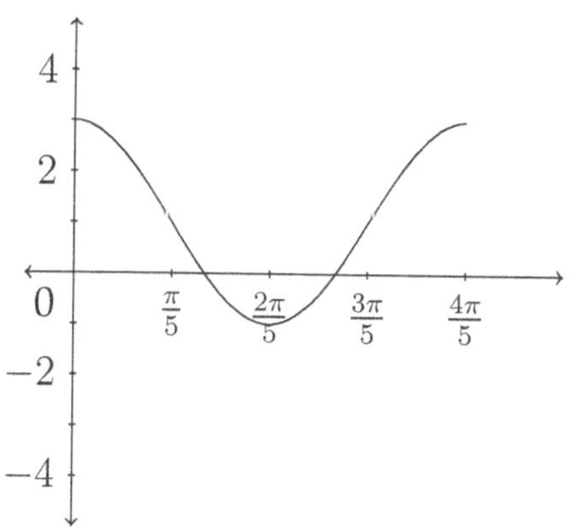

16.

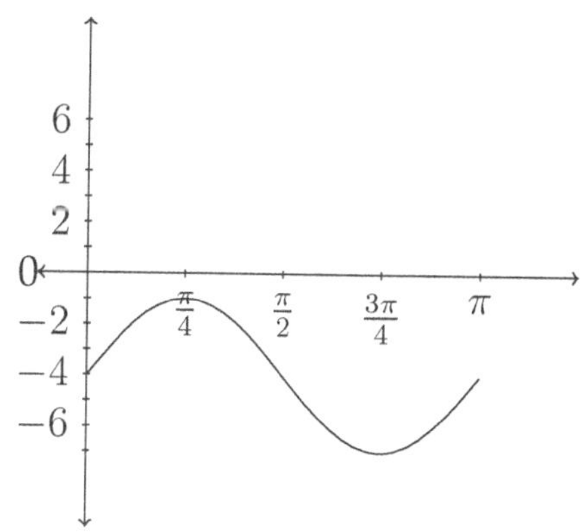

17. 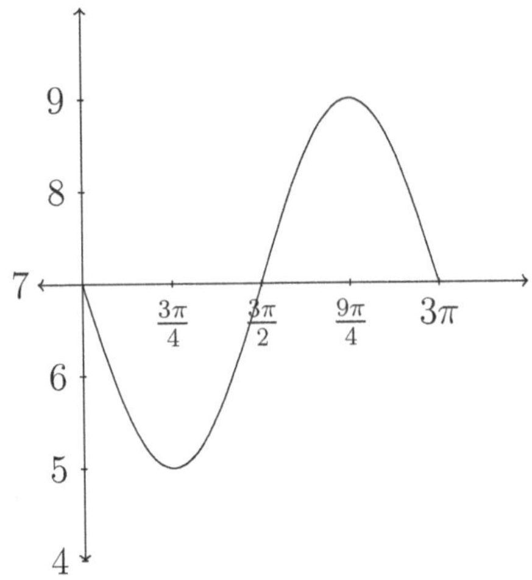
9
8
7
6
5
4
$\frac{3\pi}{4}$
$\frac{3\pi}{2}$
$\frac{9\pi}{4}$
3π

18. 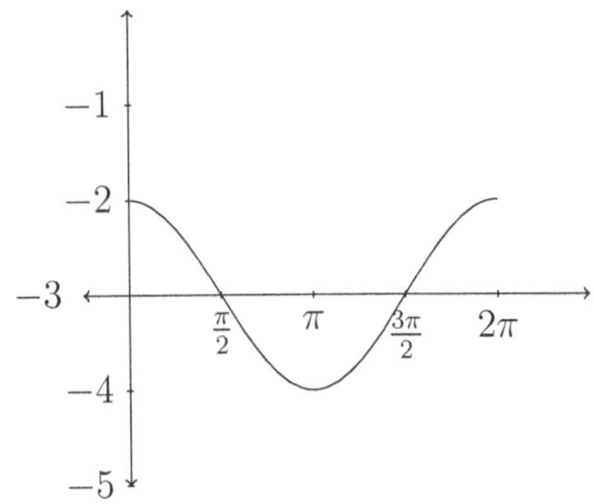
−1
−2
−3
−4
−5
$\frac{\pi}{2}$
π
$\frac{3\pi}{2}$
2π

19. 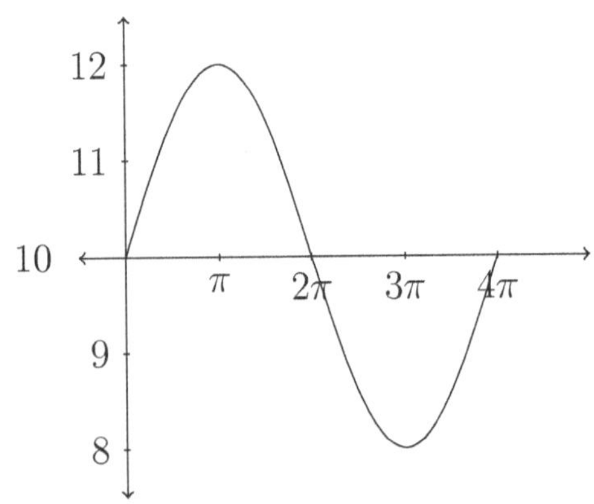
12
11
10
9
8
π
2π
3π
4π

20. 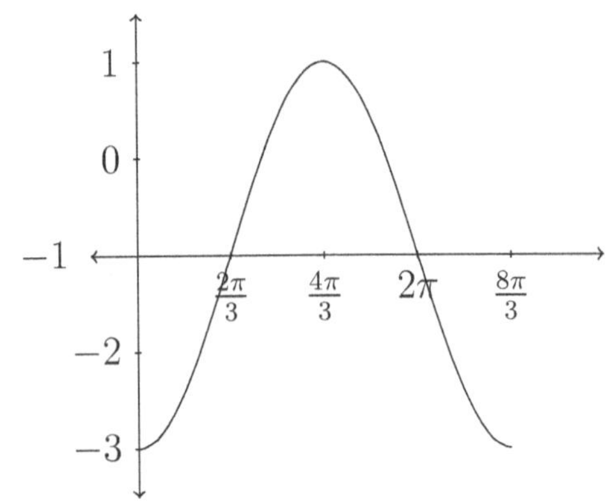
1
0
−1
−2
−3
$\frac{2\pi}{3}$
$\frac{4\pi}{3}$
2π
$\frac{8\pi}{3}$

2.4 Phase Shift

The last form of transformation we will discuss in the graphing of trigonometric functions is the phase shift, or horizontal displacement. So far, we have considered the amplitude, period and vertical shift transformations of trigonometric functions. In the standard equation $y = A\sin(Bx) + D$, these corrrespond to the coefficients A, B and D. Notice that the amplitude and vertical shift coefficients (A and D), which affect the y-axis occur outside of the trigonometric function, whereas the coefficient that affects the period of the graph along the x-axis occurs within the sine function. This is true of the phase shift as well.

If we consider a general equation of:

$$y = A\sin(Bx + C) + D$$

the constant C will affect the phase shift, or horizontal displacement of the function. Let's look at a simple example.

Example 1

Graph at least one period of the given function: $y = \sin(x + \pi)$.
Be sure to indicate important points along the x and y axes.

Let's examine this function by looking at a table of values.

x	$x+\pi$	$\sin(x+\pi)$
0	π	0
$\pi/2$	$3\pi/2$	-1
π	2π	0
$3\pi/2$	$5\pi/2$	1
2π	3π	0

Now let's look at a graph of $y = \sin(x + \pi)$ as compared to the standard graph of $y = \sin x$.

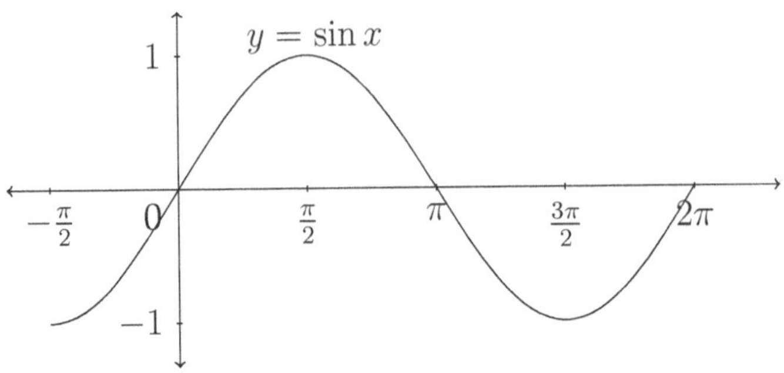

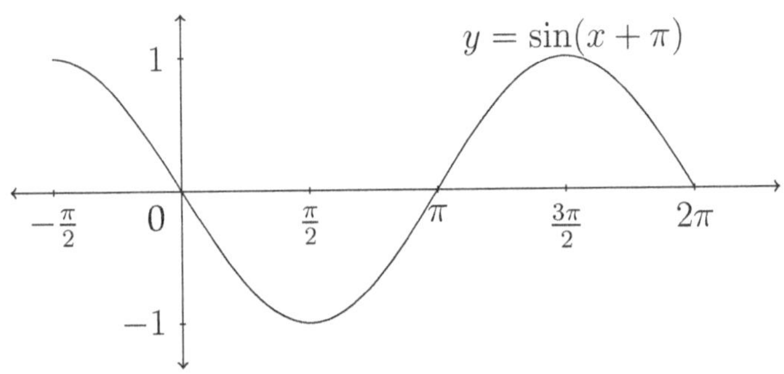

Notice that if we take the standard graph of $y = \sin x$ and drag it backwards along the x-axis a distance of π, we would have the graph of $y = \sin(x + \pi)$. That's because each x value is having π added to it, so to arrive at the x value that produces a particular y-value, we would need to subtract π. Here's an example:

$x + \pi$	$y = \sin(x + \pi)$
0	0
$\pi/2$	1
π	0
$3\pi/2$	-1
2π	0

In the table above we see the standard x and y values for the graph of the sine function. In the table below, we add a column that shows the value that x would need to be for $x + \pi$ to be the standard values:

x	$x+\pi$	$y=\sin(x+\pi)$
$-\pi$	0	0
$-\pi/2$	$\pi/2$	1
0	π	0
$\pi/2$	$3\pi/2$	-1
π	2π	0

Here's a graph of these values:

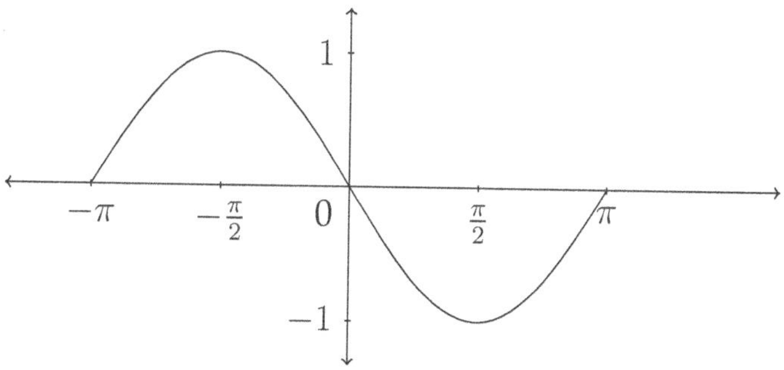

$y=\sin(x+\pi)$

This is the same graph of $y = \sin(x+\pi)$ that we saw on the previous page, but anchored to different points on the x-axis. Either graph would be a correct response to a question asking for at least one period of the graph of $y=\sin(x+\pi)$.

Let's look at another example:

Example 2

Graph at least one period of the given function: $y=\sin(x+\frac{\pi}{3})$.
Be sure to indicate important points along the x and y axes.

In this simplified example, we really have only one transformation to worry about - the phase shift. Notice that the amplitude, period and vertical shift have all been left out. When considering a sine or cosine graph that has a phase shift, a good way to start the graph of the function is to determine the new starting point of the graph. In the previous example, we saw how the function $y=\sin(x+\pi)$

shifted the graph a distance of π to the left and made the new starting point of the sine curve $-\pi$.

In graphing the standard sine curve we're generally interested in the quadrantal angles that produce the maximum, minimum and zero points of the graph. In graphing the function $y = \sin(x+ \frac{\pi}{3})$, we want to know which values of x will produce the quadrantal angles when we add $\frac{\pi}{3}$to them.

So, to determine the new starting point we want to know the solution to the equation: $x+\frac{\pi}{3}= 0$

$$x + \frac{\pi}{3} = 0$$
$$-\frac{\pi}{3} - \frac{\pi}{3}$$

$$x = -\frac{\pi}{3}$$

This is the new starting point for the graph $y = \sin(x+ \frac{\pi}{3})$. Because this graph has a standard period, the "jump" between each of the quadrantal angles will be $\frac{\pi}{2}$. To graph one period of a typical trigonometric function we'll need at least five quadrantal angle values. So, if our new starting point is $-\frac{\pi}{3}$, then the next critical value along the x-axis will be:

$$-\frac{\pi}{3} + \frac{\pi}{2} = -\frac{2\pi}{6} + \frac{3\pi}{6} = \frac{\pi}{6}$$

Then the subsequent critical values would be:

$$\frac{\pi}{6} + \frac{\pi}{2} = \frac{\pi}{6} + \frac{3\pi}{6} = \frac{4\pi}{6} = \frac{2\pi}{3}$$

$$\frac{4\pi}{6} + \frac{3\pi}{6} = \frac{7\pi}{6}$$

$$\frac{7\pi}{6} + \frac{3\pi}{6} = \frac{10\pi}{6} = \frac{5\pi}{3}$$

So the five critical values along the x-axis are:

$$-\frac{2\pi}{6}, \frac{\pi}{6}, \frac{4\pi}{6}, \frac{7\pi}{6} \text{and } \frac{10\pi}{6}$$

or, in reduced form:

$$-\frac{\pi}{3}, \frac{\pi}{6}, \frac{2\pi}{3}, \frac{7\pi}{6} \text{and } \frac{5\pi}{3}$$

In order to graph the function, we would put these values along the x-axis and plot the standard quadrantal y-values to match up with them:

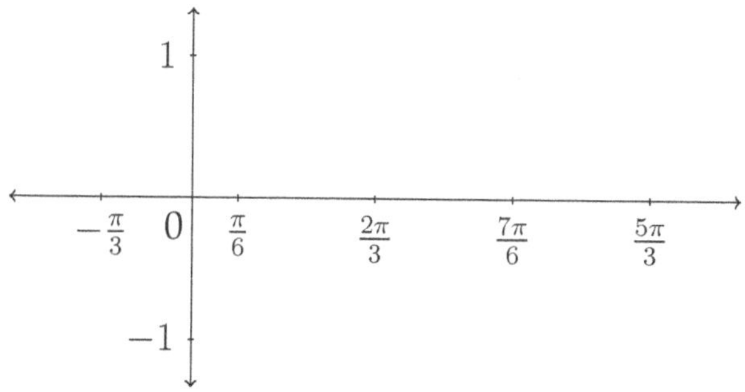

The y-values for the sine function start at zero, go up to the maximum, back down through zero to the minimum and then back to zero:

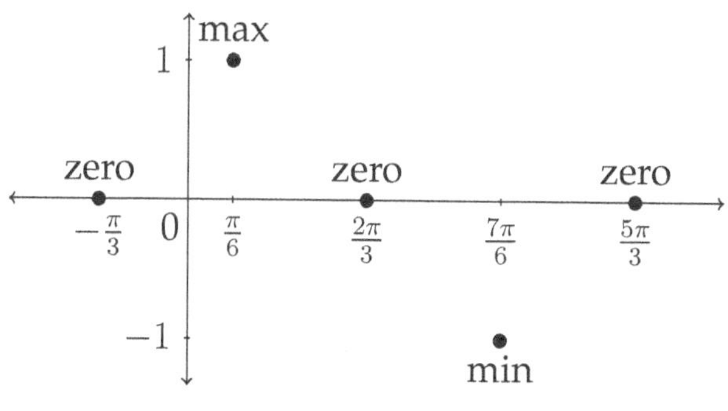

Connecting these points to make a sine curve produces the following graph:

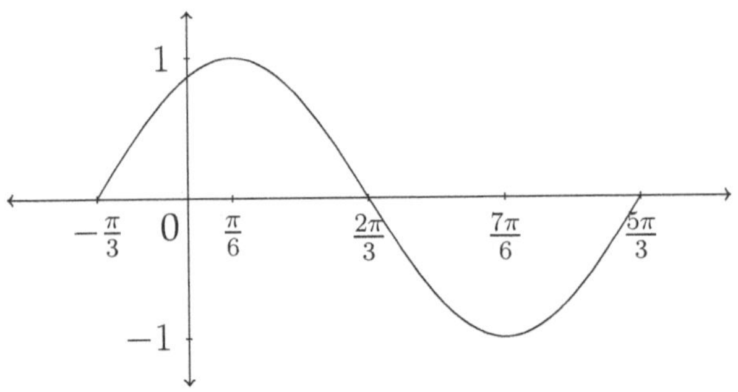

Exercises 2.4

Match each function with the appropriate graph.

1. $y = \cos(x - \frac{\pi}{4})$
2. $y = \sin(x + \frac{\pi}{4})$
3. $y = \cos x - 1$
4. $y = \sin x + 1$
5. $y = \sin(x - \frac{\pi}{4})$
6. $y = 1 - \cos x$
7. $y = \sin x - 1$
8. $y = \cos(x + \frac{\pi}{4})$

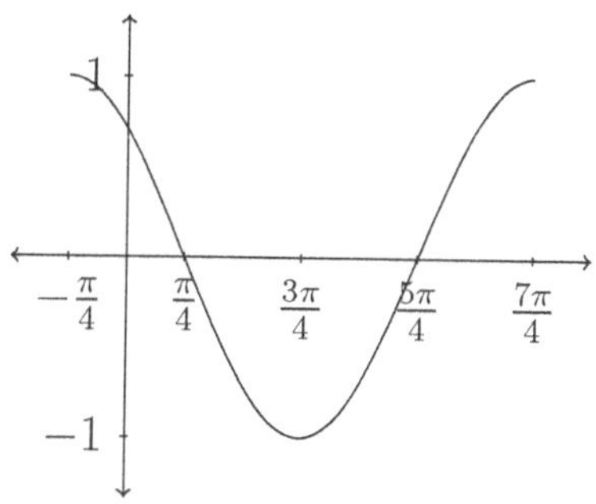

A.

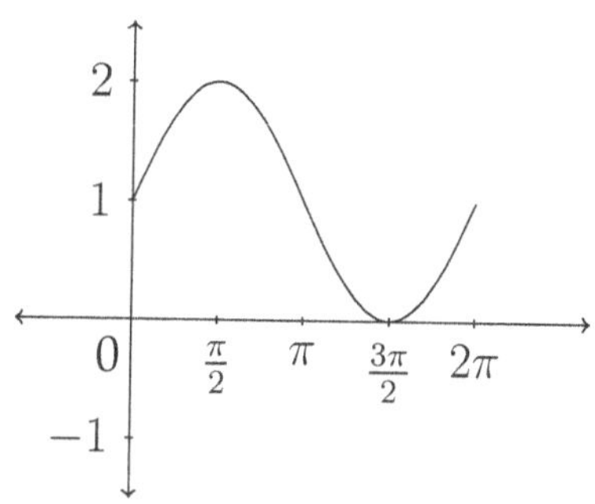

B.

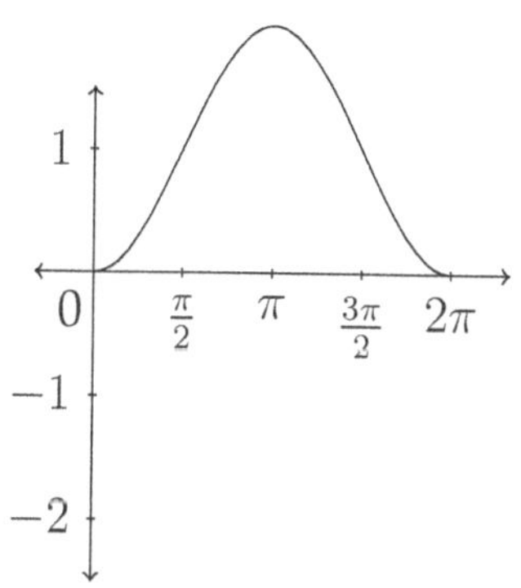

C.

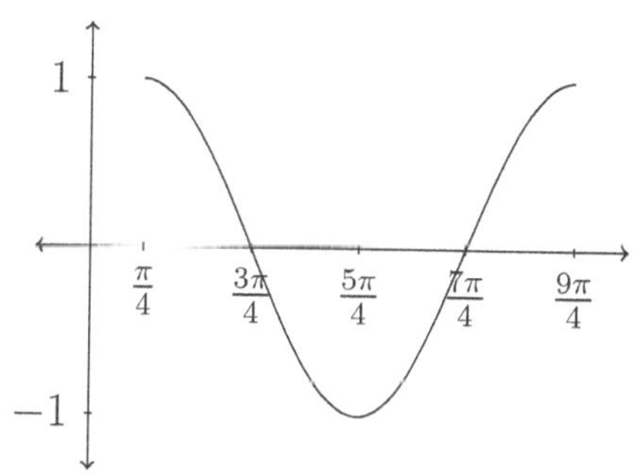

D.

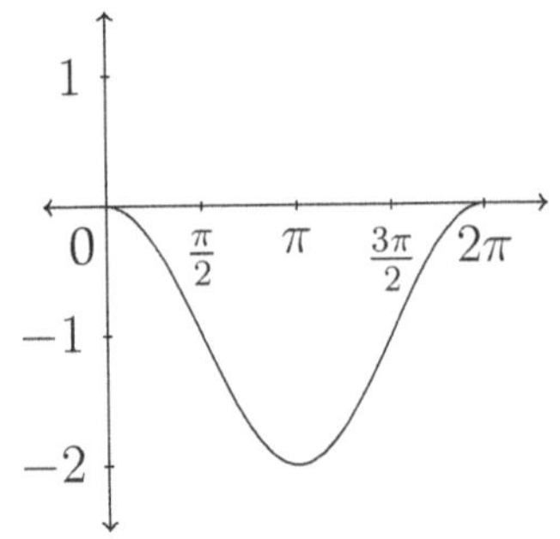

E.

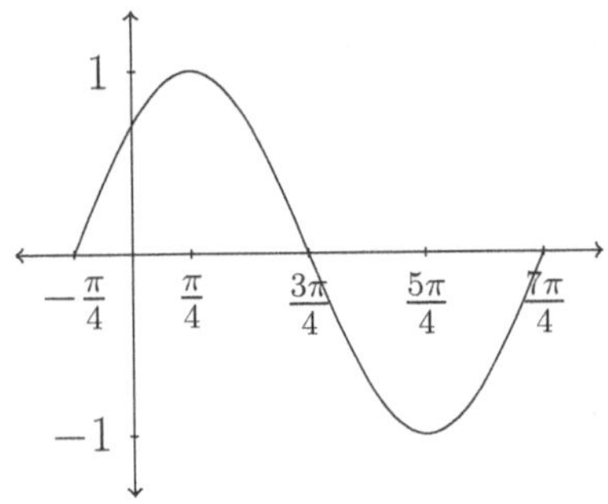

F.

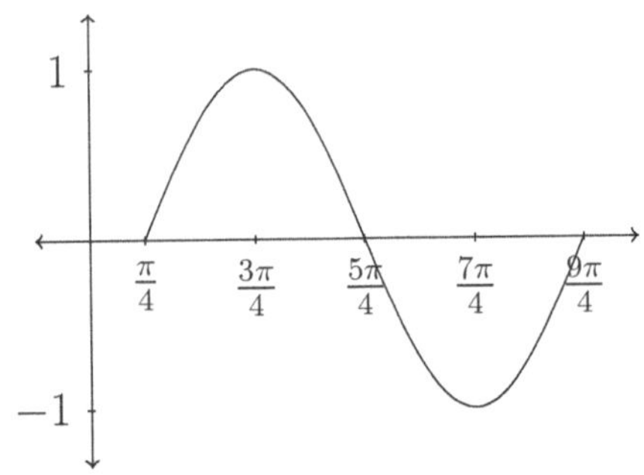

G.

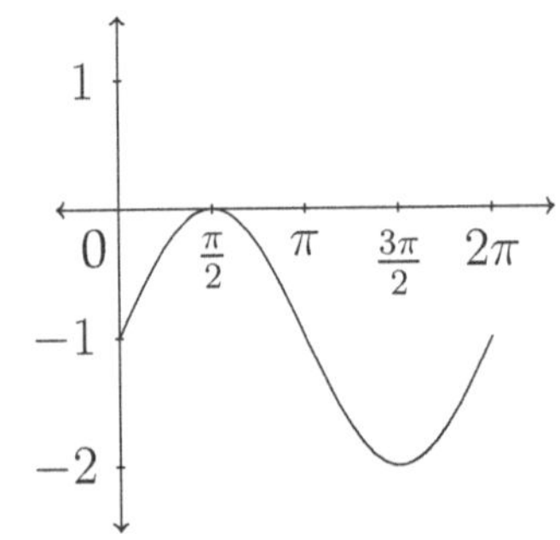

H.

Sketch at least one period for each function. Be sure to include the important values along the x and y axes.

9. $y = \sin(x + \frac{\pi}{6})$

10. $y = \cos(x - \frac{\pi}{6})$

11. $y = \cos(x - \frac{\pi}{3})$

12. $y = \sin(x + \frac{\pi}{3})$

13. $y = \sin(x - \frac{3\pi}{4})$

14. $y = \cos(x + \frac{3\pi}{4})$

15. $y = \cos(x + \frac{2\pi}{3})$

16. $y = \sin(x - \frac{2\pi}{3})$

2.5 Combining the Transformations

In the previous sections, we have seen how the various transformations act on the trigonometric functions and we have worked with the first three (amplitude, period and vertical shift) in combination with each other. Combining the phase shift with the other transformations is tricky because of the way that the period and the phase shift interact with each other.

Now we have two standard equations for the sinusoid:

$$y = A\sin(Bx + C) + D$$
and
$$y = A\cos(Bx + C) + D$$

A and D, the amplitude and the vertical shift affect the y-axis, while B and C affect the x-axis.

y-axis	x-axis
Amplitude= $\|A\|$	Period=$\frac{2\pi}{B}$
Vertical Shift= D	Phase Shift=$-\frac{C}{B}$

Let's look at an example in which we need to combine a change in the period of the graph with a phase shift.

Example 1

Graph at least one period of the given function. Indicate the important values along the x and y axes.

$y = \cos(4x + \pi)$

The transformations in this example only affect the x-axis. The period of the function is $\frac{2\pi}{B} = \frac{2\pi}{4} = \frac{\pi}{2}$. So, the function will complete one full cycle over a distance of $\frac{\pi}{2}$along the x-axis.

However, because of the phase shift, this graph *will not* start at 0 and end at $\frac{\pi}{2}$. We need to find the new starting point that is caused by the phase shift. So, we take what is called the "argument," or what it is we're finding the cosine of: $(4x + \pi)$ and set that equal to zero.

$$4x + \pi = 0$$
$$4x = -\pi$$
$$x = -\frac{\pi}{4}$$

This is our new starting point. To identify the critical values along the x-axis, we'll need to determine how far each "jump" would be given a period of $\frac{\pi}{2}$.

$$\frac{\pi}{2} * \frac{1}{4} = \frac{\pi}{8}$$

So, each subsequent critical value along the x-axis will be a distance of $\frac{\pi}{8}$ from the previous one. If we start at our new starting point for this function $-\frac{\pi}{4}$, then if we add $\frac{\pi}{8}$ a total of 4 times, we will arrive at each of the five critical values for this function.

$$-\frac{\pi}{4} + \frac{\pi}{8} = -\frac{2\pi}{8} + \frac{\pi}{8} = -\frac{\pi}{8}$$

$$-\frac{\pi}{8} + \frac{\pi}{8} = 0$$

$$0 + \frac{\pi}{8} = \frac{\pi}{8}$$

$$\frac{\pi}{8} + \frac{\pi}{8} = \frac{2\pi}{8} = \frac{\pi}{4}$$

So the critical values along the x-axis would be:

$$-\frac{\pi}{4}, -\frac{\pi}{8}, 0, \frac{\pi}{8}, \text{and } \frac{\pi}{4}$$

Notice that the distance between the starting point $-\frac{\pi}{4}$ and the ending point $\frac{\pi}{4}$ is equal to the period we found at the beginning of the problem, which was $\frac{\pi}{2}$. Now let's graph the function:

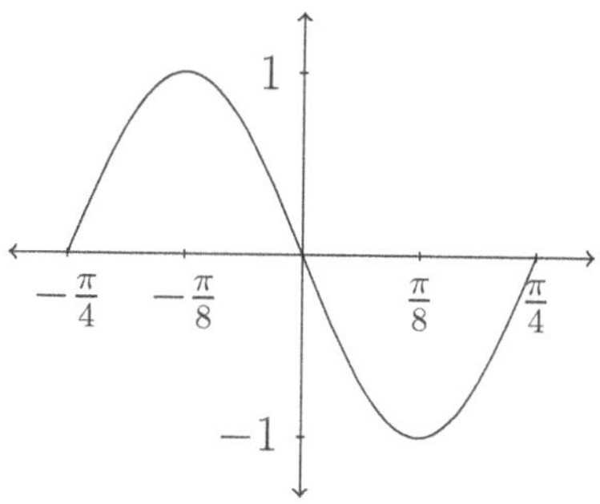

Since there were no changes to the y-axis, the amplitude for the function is 1 and the vertical shift is 0. Along the x-axis, we see a positive sine function that starts at $(-\frac{\pi}{4}, 0)$ then goes up to $(-\frac{\pi}{8}, 1)$, back down through $(0, 0)$ to $(\frac{\pi}{8}, -1)$ and back up to $(\frac{\pi}{4}, 0)$ to complete one full cycle of the graph.

Let's look at an example in which there are some changes to the y-axis as well as the x-axis.

Example 2

Graph at least one period of the given function. Be sure to identify critical values along the x and y axes.

$y = -\frac{5}{2} + \cos(3x - \pi)$

Remember which coefficients affect which axis in graphing:

y-axis	x-axis
Amplitude= $\lvert A\rvert$	Period= $\frac{2\pi}{B}$
Vertical Shift= D	Phase Shift= $-\frac{C}{B}$

In this example, the amplitude is 1, since there is no coefficient in front of the cosine function. The vertical shift is $-\frac{5}{2}$, which will shift the function down a distance of 2.5 on the y-axis. So, the mid-line or "zero" points of the graph will be at -2.5, the maximum y value will be -1.5 and the minimum y value will be -3.5.

Along the x-axis, the period for the graph will be $\frac{2\pi}{B}=\frac{2\pi}{3}$, since the coefficient B in this problem is 3. To find the new starting point, we'll take the argument of the cosine function and set it equal to zero.

$$3x - \pi = 0$$

$$3x = \pi$$

$$x = \pi * \frac{1}{3} = \frac{\pi}{3}$$

So, our new starting point will be at $\frac{\pi}{3}$. To determine the other critical values along the x-axis, we can find out how far each "jump" between the critical values would be. To do this, we take the period $(\frac{2\pi}{3})$ and divide it by 4 (or multiply by $\frac{1}{4}$).

$$\frac{2\pi}{3} * \frac{1}{4} = \frac{2\pi}{12} = \frac{\pi}{6}$$

Now we can add this value to our new starting point four times to determine the other critical values along the x-axis.

$$\frac{\pi}{3} + \frac{\pi}{6} = \frac{2\pi}{6} + \frac{\pi}{6} = \frac{3\pi}{6} = \frac{\pi}{2}$$

$$\frac{3\pi}{6} + \frac{\pi}{6} = \frac{4\pi}{6} = \frac{2\pi}{3}$$

$$\frac{4\pi}{6} + \frac{\pi}{6} = \frac{5\pi}{6}$$

$$\frac{5\pi}{6} + \frac{\pi}{6} = \pi$$

So the critical values along the x-axis would be:

$$\frac{2\pi}{6}, \frac{3\pi}{6}, \frac{4\pi}{6}, \frac{5\pi}{6}, \text{and } \frac{6\pi}{6}$$

or

$$\frac{\pi}{3}, \frac{\pi}{2}, \frac{2\pi}{3}, \frac{5\pi}{6}, \text{and } \pi$$

Again, notice that the distance along the x-axis from the starting point to the ending point is the period: $\frac{2\pi}{3}$. Now let's graph the function:

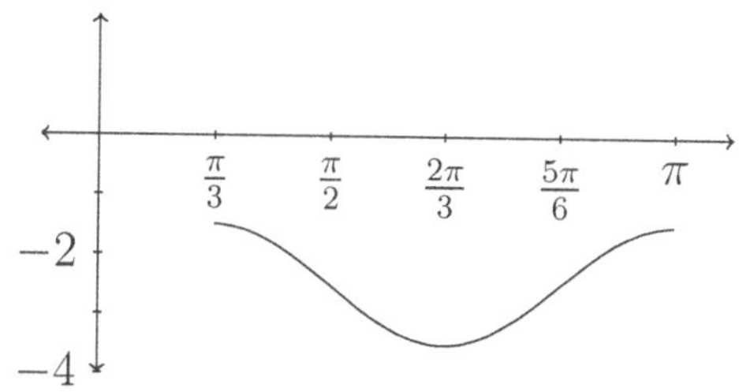

or

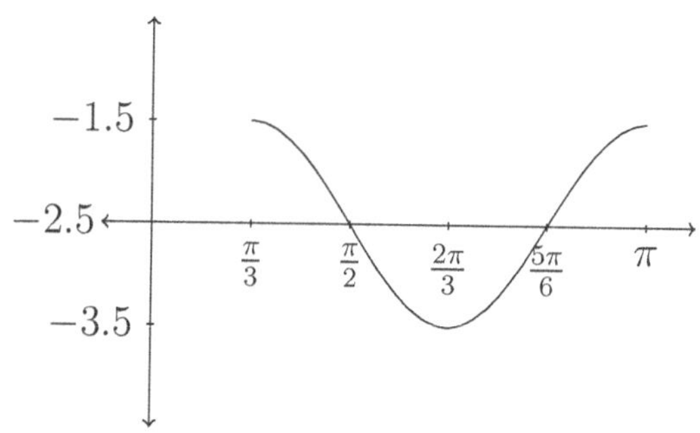

Let's look at one more example.

Example 3

Sometimes the coefficient B appears factored out of the argument as it does in the problem below.

Graph at least one period of the given function. Be sure to include the critical values along the x and y axes.

$y = 4\sin 2(x+\frac{\pi}{3}) - 1.$

First let's see how the x and y axes are affected by the transformations in this problem.

<u>y-axis</u>	<u>x-axis</u>
Amplitude= $\|A\|$	Period=$\frac{2\pi}{B}$
Vertical Shift= D	Phase Shift=$-\frac{C}{B}$

$y = 4\sin 2(x+\frac{\pi}{3}) - 1.$

The amplitude in this problem is 4 and the vertical shift is -1.

The period for this graph is $\frac{2\pi}{B}=\frac{2\pi}{2}= \pi$. Notice that the value of B is 2 in this example, even though it's been factored out from the rest of the argument.

The new starting point for the graph is actually easier to find in problems of this type. If we take the argument as it is and set it equal to zero:

$$2(x+\frac{\pi}{3}) = 0$$

we can divide through on both sides by 2 to cancel out the factor of B:

$$\frac{2(x+\frac{\pi}{3})}{2}=\frac{0}{2}$$

$$x+\frac{\pi}{3}= 0$$

$$x =-\frac{\pi}{3}$$

So, the new starting point for the function is $-\frac{\pi}{3}$.

Now let's find the rest of the critical values along the x-axis. The period for this graph is π, so the "jump" between the critical values along the x-axis will be:

$$\pi*\frac{1}{4}=\frac{\pi}{4}$$

To find the rest of the critical values we'll need to add $\frac{\pi}{4}$to the starting point of the graph $(-\frac{\pi}{3})$ four times:

$$-\frac{\pi}{3}+\frac{\pi}{4}=-\frac{4\pi}{12}+\frac{3\pi}{12}=-\frac{\pi}{12}$$

$$-\frac{\pi}{12}+\frac{3\pi}{12}=\frac{2\pi}{12}=\frac{\pi}{6}$$

$$\frac{2\pi}{12}+\frac{3\pi}{12}=\frac{5\pi}{12}$$

$$\frac{5\pi}{12}+\frac{3\pi}{12}=\frac{8\pi}{12}=\frac{2\pi}{3}$$

So the critical values along the x-axis would be:

$$-\frac{4\pi}{12}, -\frac{1\pi}{12}, \frac{2\pi}{12}, \frac{5\pi}{12}, \text{and } \frac{8\pi}{12}$$

or

$$-\frac{\pi}{3}, -\frac{\pi}{12}, \frac{\pi}{6}, \frac{5\pi}{12}, \text{and } \frac{2\pi}{3}$$

Now that we've addressed each of the four transformations let's use this information to draw the graph. First the y-axis - the amplitude is 4 and the vertical shift is -1:

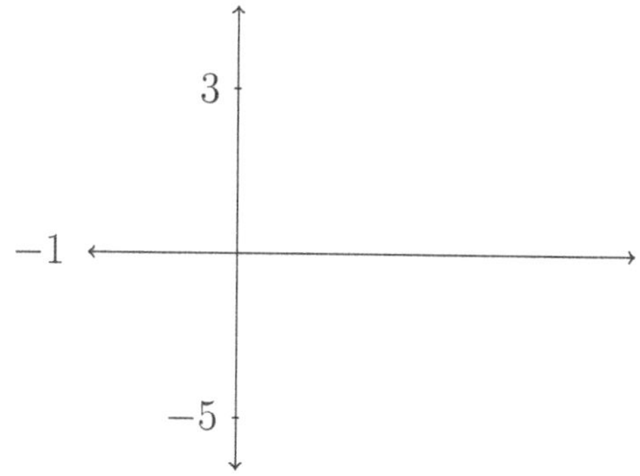

Now, let's fill in the information for the x-axis. The critical values along the x-axis are $-\frac{\pi}{3}$, $-\frac{\pi}{12}$, $\frac{\pi}{6}$, $\frac{5\pi}{12}$, and $\frac{2\pi}{3}$

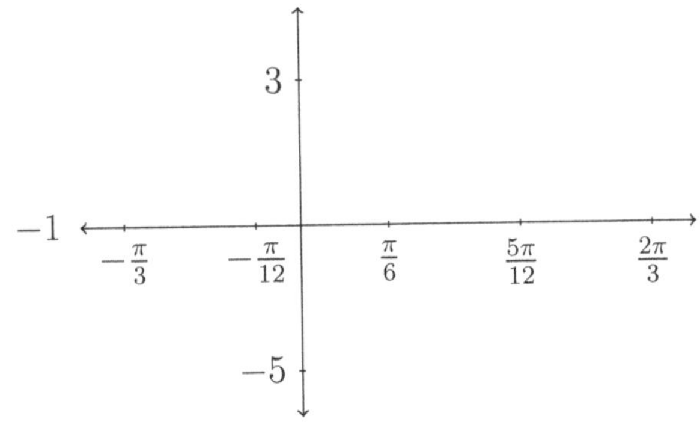

The function we're graphing is a positive sine function, so it will start at the "mid-line" or zero value (which in this case is -1), go up to the maximum, back through the mid-line to the minimum and back to the mid-line:

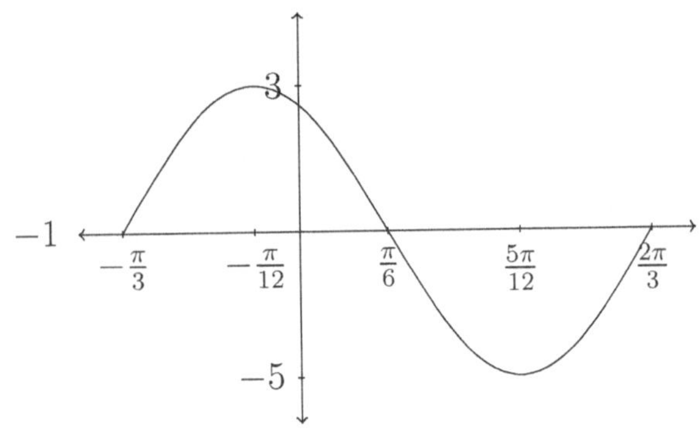

Exercises 2.5

Determine the Amplitude, Period, Vertical Shift and Phase Shift for each function and graph at least one complete period. Be sure to identify the critical values along the x and y axes.

1. $y = \sin(x+\frac{\pi}{2})$
2. $y = \sin(x - \pi)$
3. $y = 3\cos(x-\frac{\pi}{2})$
4. $y = \frac{1}{2}\cos(x + \pi)$
5. $y = 3 + \cos(x-\frac{\pi}{4})$
6. $y = -2 + \sin(x+\frac{\pi}{6})$
7. $y = \sin(2x - \pi)$
8. $y = \sin(4x+\frac{\pi}{4})$
9. $y = 2\cos(\frac{x}{2}+\pi)$
10. $y = -3\sin(6x - \pi)$
11. $y = -\frac{1}{3}\sin(2x+\frac{\pi}{4})$
12. $y = \frac{1}{2}\cos(\frac{x}{2}-\pi)$
13. $y = 2\sin(2x-\frac{\pi}{3}) - 1$
14. $y = 1 + 2\cos(3x+\frac{\pi}{2})$
15. $y = 3\cos 2(x+\frac{\pi}{6})$
16. $y = -4\sin 2(x+\frac{\pi}{2})$
17. $y = \sin\frac{1}{2}(x+\frac{\pi}{4})$
18. $y = 3 + 2\sin 3(x+\frac{\pi}{2})$

In problems $19 - 22$, determine an equation for the function that is shown.

19.

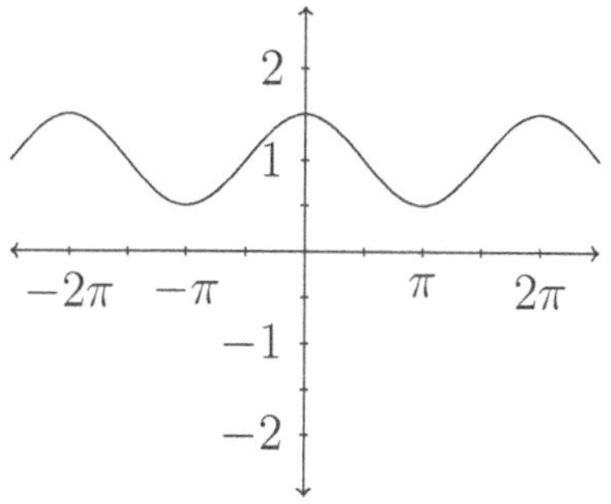

20.

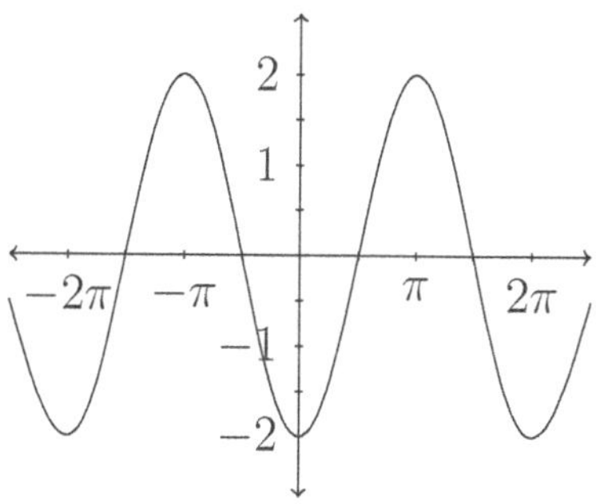

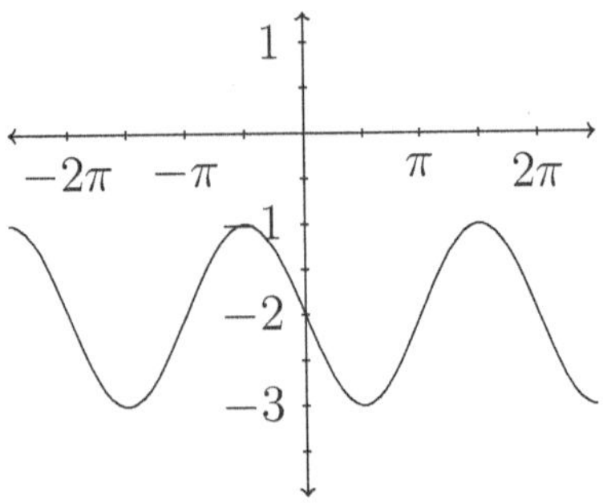

21.

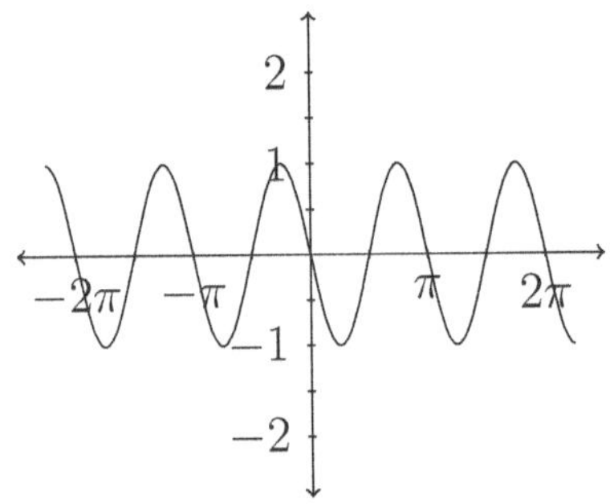

22.

Match the function to the appropriate graph

23. $y = -\cos 2x$

24. $y = \frac{1}{2}\sin x - 2$

25. $y = 2\cos(x + \frac{\pi}{2})$

26. $y = -3\sin\frac{1}{2}x - 1$

27. $y = \sin(x - \pi) - 2$

28. $y = -\frac{1}{2}\cos(x - \frac{\pi}{4})$

29. $y = \frac{1}{3}\sin 3x$

30. $y = \cos(x - \frac{\pi}{2})$

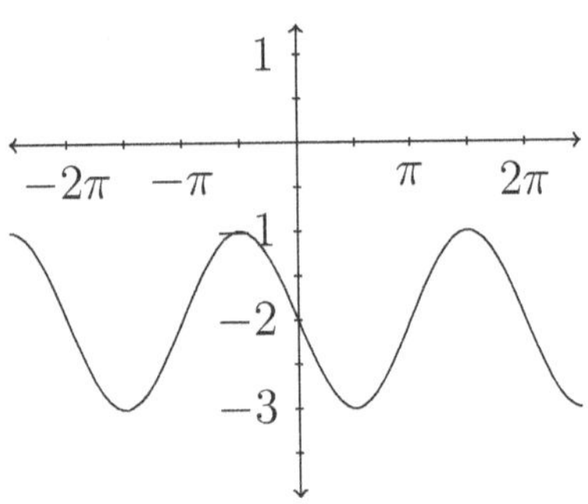

A.

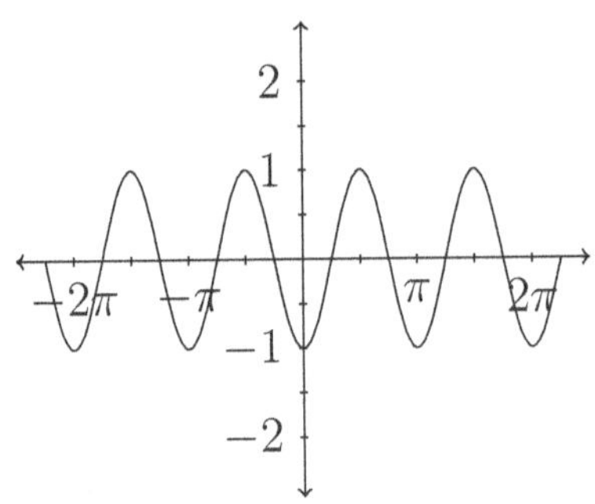

B.

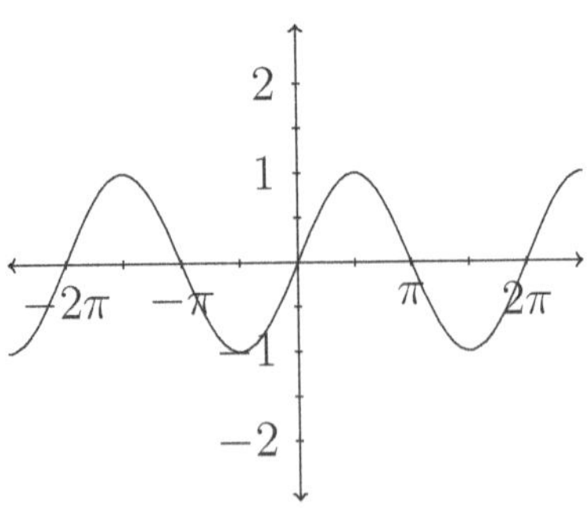

C.

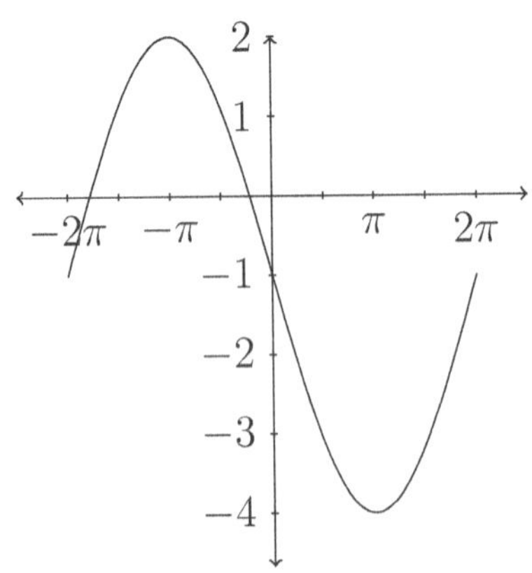

D.

E. 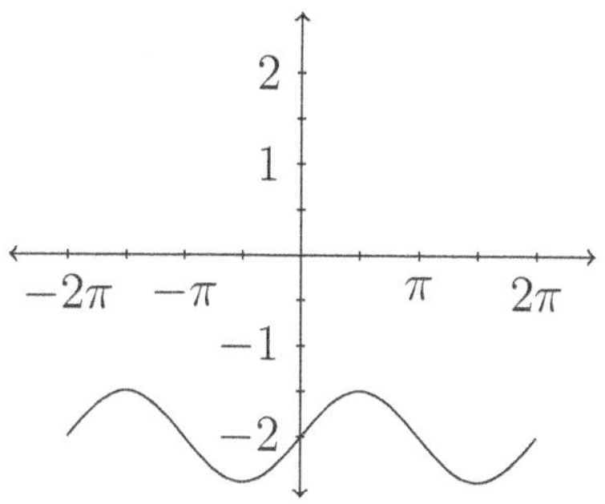

F. 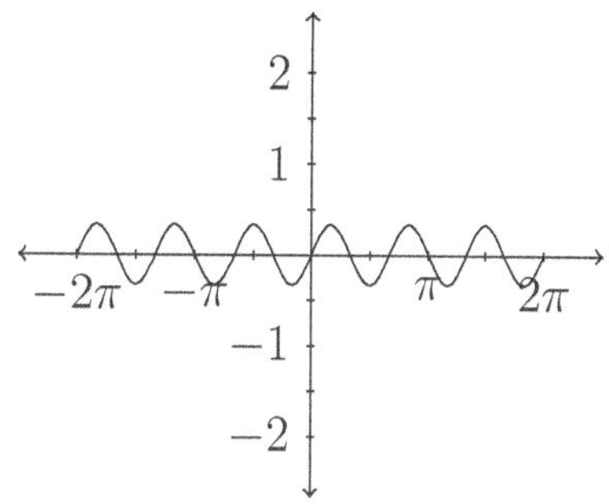

G. 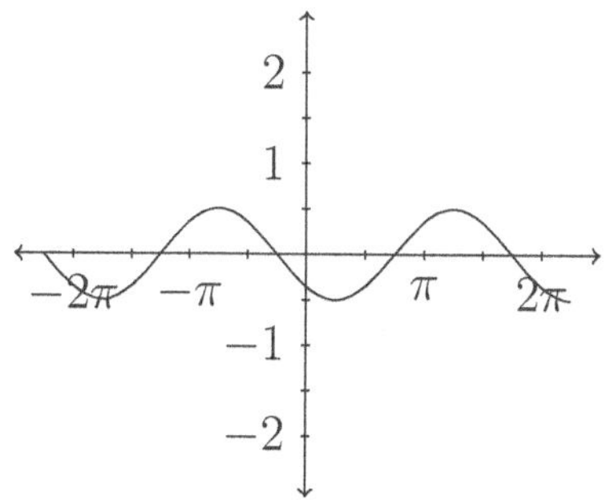

H. 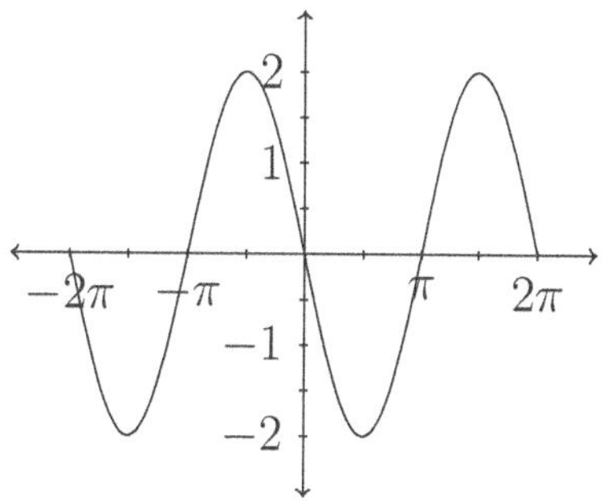

Determine the Amplitude, Period, Vertical Shift and Phase Shift for each function and graph at least one complete period. Be sure to identify the critical values along the x and y axes.

31. $y = 2\cos(2x + \frac{\pi}{2}) - 1$

32. $y = -4\cos(3x - 2\pi)$

33. $y = \sin(2x - \frac{\pi}{4})$

34. $y = -\sin(3x + \pi)$

35. $y = 3\cos(x + \frac{\pi}{3}) + 1$

36. $y = -2\sin(3x - \frac{\pi}{2}) + 4$

37. $y = -\frac{1}{2}\sin(x - \frac{\pi}{2}) - 2$

38. $y = 2 - \cos(2x - \frac{\pi}{3})$

Chapter 3

Trigonometric Identities and Equations

Due to the nature of the trigonometric ratios, they have some interesting properties that make them useful in a number of mathematical problem-solving situations. One of the hallmarks of mathematical problem-solving is to change the appearance of the problem without changing its value. Trigonometric identities can be very helpful in changing the appearance of a problem.

The process of demonstrating the validity of a trigonometric identity involves changing one trigonometric expression into another, using a series of clearly defined steps. We'll look at a few examples briefly, but first, let's examine some of the fundamental trigonometric identities.

3.1 Reciprocal and Pythagorean Identities

The two most basic types of trigonometric identities are the reciprocal identities and the Pythagorean identities. The reciprocal identities are simply definitions of the reciprocals of the three standard trigonometric ratios:

$$\sec\theta = \frac{1}{\cos\theta} \qquad \csc\theta = \frac{1}{\sin\theta} \qquad \cot\theta = \frac{1}{\tan\theta}$$

Also, recall the definitions of the three standard trigonometric ratios (sine, cosine and tangent):

$$\sin\theta = \frac{opp}{hyp}$$

$$\cos\theta = \frac{adj}{hyp}$$

$$\tan\theta = \frac{opp}{adj}$$

If we look more closely at the relationships between the sine, cosine and tangent, we'll notice that $\frac{\sin\theta}{\cos\theta} = \tan\theta$.

$$\frac{\sin\theta}{\cos\theta} = \frac{\left(\frac{opp}{hyp}\right)}{\left(\frac{adj}{hyp}\right)} = \frac{opp}{hyp} * \frac{hyp}{adj} = \frac{opp}{adj} = \tan\theta$$

Pythagorean Identities

The Pythagorean Identities are, of course, based on the Pythagorean Theorem. If we recall a diagram that was introduced in Chapter 2, we can build these identities from the relationships in the diagram:

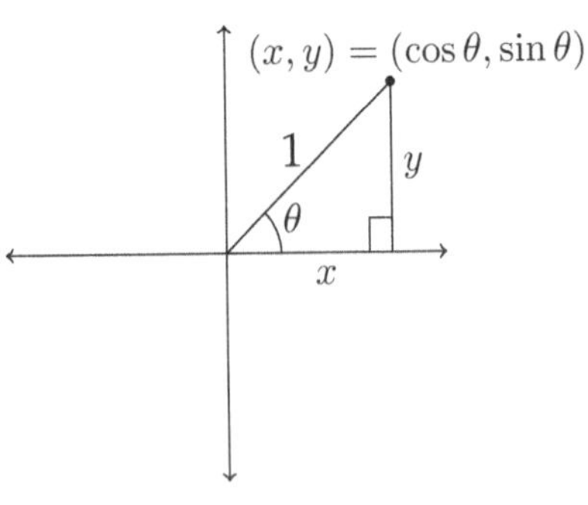

Using the Pythagorean Theorem in this diagram, we see that $x^2 + y^2 = 1^2$, so $x^2 + y^2 = 1$. But, also remember that, in the unit circle, $x = \cos\theta$ and $y = \sin\theta$.

Substituting this equality gives us the first Pythagorean Identity:

$$x^2 + y^2 = 1$$
or
$$\cos^2\theta + \sin^2\theta = 1$$

This identity is usually stated in the form:

$$\sin^2\theta + \cos^2\theta = 1$$

If we take this identity and divide through on both sides by $\cos^2\theta$, this will result in the first of two additional Pythagorean Identities:

$$\frac{\sin^2\theta}{\cos^2\theta} + \frac{\cos^2\theta}{\cos^2\theta} = \frac{1}{\cos^2\theta}$$

or

$$\tan^2\theta + 1 = \sec^2\theta$$

Dividing through by $\sin^2\theta$ gives us the second:

$$\frac{\sin^2\theta}{\sin^2\theta} + \frac{\cos^2\theta}{\sin^2\theta} = \frac{1}{\sin^2\theta}$$

or

$$1 + \cot^2\theta = \csc^2\theta$$

So, the three Pythagorean Identities we will be using are:

$$\sin^2\theta + \cos^2\theta = 1$$

$$\tan^2\theta + 1 = \sec^2\theta$$

$$1 + \cot^2\theta = \csc^2\theta$$

These Pythagorean Identities are often stated in other terms, such as:

$$\sin^2\theta = 1 - \cos^2\theta$$

$$\cos^2\theta = 1 - \sin^2\theta$$

$$\tan^2\theta = \sec^2\theta - 1$$

$$\cot^2\theta = \csc^2\theta - 1$$

At the beginning of this chapter, we discussed verifying trigonometric identities. Now that we have some basic identities to work with, let's use them to verify the equality of some more complicated statements. The process of verifying trigonometric identities involves changing one side of the given expression into the other side. Since these are not really equations, we will not treat them the way we treat equations. That is to say, we won't add or subtract anything to both sides of the statement (or multiply or divide by anything on both sides either).

Another reason for not treating a trigonometric identity as an equation is that, in practice, this process typically involves just one side of the statement. In problem solving, mathematicians typically use trigonometric identities to change the appearance of a problem without changing its value. In this process, a trigonometric expression is changed into another trigonometric expression rather than showing that two trigonometric expressions are the same, which is what we will be doing.

Example 1

Verify the identity $(\sin\theta)(\cot\theta) = \cos\theta$

This is a very straightforward identity and it can solved by using one of the fundamental approaches to working with trigonometric identities. This is the approach of writing everything in terms of sines and cosines.

Beginning with the original statement:

$$(\sin\theta)(\cot\theta) = \cos\theta$$

Replace $\cot\theta$ with $\frac{\cos\theta}{\sin\theta}$:

$$(\sin\theta)\frac{\cos\theta}{\sin\theta} = \cos\theta$$

Then canceling out the $\sin\theta$:

$$\cos\theta = \cos\theta$$

There are four fundamental approaches to verifying trigonometric identities:

1. write everything in terms of sines and cosines
2. make a common denominator and add fractions
3. split a fraction
4. factor and cancel

Not all of these can be used in every problem and some problems will use combinations of these strategies. Here is another example.

Example 2

Verify the identity $\tan\theta + \cot\theta = \sec\theta\csc\theta$.

First we'll write everything in terms of sines and cosines:

$$\tan\theta + \cot\theta = \sec\theta\csc\theta$$

$$\frac{\sin\theta}{\cos\theta} + \frac{\cos\theta}{\sin\theta} = \frac{1}{\cos\theta}\cdot\frac{1}{\sin\theta}$$

Next, on the left hand side, we can add the two fractions together by making a common denominator of $\cos\theta\sin\theta$.

$$\frac{\sin\theta}{\cos\theta}+\frac{\cos\theta}{\sin\theta}=\frac{1}{\cos\theta}\cdot\frac{1}{\sin\theta}$$

$$\frac{\sin\theta}{\sin\theta}\cdot\frac{\sin\theta}{\cos\theta}+\frac{\cos\theta}{\sin\theta}\cdot\frac{\cos\theta}{\cos\theta}=\frac{1}{\cos\theta}\cdot\frac{1}{\sin\theta}$$

$$\frac{\sin^2\theta}{\sin\theta\cos\theta}+\frac{\cos^2\theta}{\sin\theta\cos\theta}=\frac{1}{\cos\theta}\cdot\frac{1}{\sin\theta}$$

$$\frac{\sin^2\theta+\cos^2\theta}{\sin\theta\cos\theta}=\frac{1}{\cos\theta}\cdot\frac{1}{\sin\theta}$$

$$\frac{1}{\sin\theta\cos\theta}=\frac{1}{\sin\theta\cos\theta}$$

In this example, you can see that we have first written everything in terms of sines and cosines, then created common denominators and added the fractions on the left hand side together. After this is done, we can replace the expression $\sin^2\theta+\cos^2\theta$ with 1, since this is the fundamental Pythagorean Identity.

Example 3

Verify the identity $\frac{\tan\theta-\cot\theta}{\sin\theta\cos\theta}=\sec^2\theta-\csc^2\theta$

We'll begin this problem by splitting the fraction over the denominator. This can be helpful in problems in which there is no addition or subtraction in the denominator. The idea here is that since $\frac{a}{x}+\frac{b}{x}=\frac{a+b}{x}$, then we can reverse this process and say that $\frac{a+b}{x}=\frac{a}{x}+\frac{b}{x}$.

In the problem above we'll say that:

$$\frac{\tan\theta-\cot\theta}{\sin\theta\cos\theta}=\sec^2\theta-\csc^2\theta$$

$$\frac{\tan\theta}{\sin\theta\cos\theta}-\frac{\cot\theta}{\sin\theta\cos\theta}=\sec^2\theta-\csc^2\theta$$

$$\frac{\frac{\sin\theta}{\cos\theta}}{\sin\theta\cos\theta}-\frac{\frac{\cos\theta}{\sin\theta}}{\sin\theta\cos\theta}=\sec^2\theta-\csc^2\theta$$

$$\frac{\sin\theta}{\cos\theta}\cdot\frac{1}{\sin\theta\cos\theta}-\frac{\cos\theta}{\sin\theta}\cdot\frac{1}{\sin\theta\cos\theta}=\sec^2\theta-\csc^2\theta$$

$$\frac{\cancel{\sin\theta}}{\cos\theta}\cdot\frac{1}{\cancel{\sin\theta}\cos\theta}-\frac{\cancel{\cos\theta}}{\sin\theta}\cdot\frac{1}{\sin\theta\cancel{\cos\theta}}=\sec^2\theta-\csc^2\theta$$

$$\frac{1}{\cos^2\theta}-\frac{1}{\sin^2\theta}=\sec^2\theta-\csc^2\theta$$

$$\sec^2\theta-\csc^2\theta=\sec^2\theta-\csc^2\theta$$

Example 4

Verify the identity $\frac{\tan^2\theta-\cos^2\theta}{1-\cos^2\theta}=\sec^2\theta-\cot^2\theta$.

On the left-hand side, notice the expression $1-\cos^2\theta$ in the denominator. We can replace this with $\sin^2\theta$, which is a simpler expression. It is often helpful to have a simpler expression in the denominator rather than a more complicated expression.

$$\frac{\tan^2\theta - \cos^2\theta}{1 - \cos^2\theta} = \sec^2\theta - \cot^2\theta$$

$$\frac{\tan^2\theta - \cos^2\theta}{\sin^2\theta} = \sec^2\theta - \cot^2\theta$$

Next, we can split the fraction over the denominator of $\sin^2\theta$:

$$\frac{\tan^2\theta - \cos^2\theta}{\sin^2\theta} = \sec^2\theta - \cot^2\theta$$

$$\frac{\tan^2\theta}{\sin^2\theta} - \frac{\cos^2\theta}{\sin^2\theta} = \sec^2\theta - \cot^2\theta$$

We can see on the left-hand side that the expression $\frac{\cos^2\theta}{\sin^2\theta}$ is equivalent to $\cot^2\theta$, but the first piece on the left-hand side needs to be simplified a little more. We'll rewrite $\tan^2\theta$ as $\frac{\sin^2\theta}{\cos^2\theta}$ and then simplify the complex fraction.

$$\frac{\tan^2\theta}{\sin^2\theta} - \frac{\cos^2\theta}{\sin^2\theta} = \sec^2\theta - \cot^2\theta$$

$$\frac{\frac{\sin^2\theta}{\cos^2\theta}}{\sin^2\theta} - \cot^2\theta = \sec^2\theta - \cot^2\theta$$

$$\frac{\sin^2\theta}{\cos^2\theta} \cdot \frac{1}{\sin^2\theta} - \cot^2\theta = \sec^2\theta - \cot^2\theta$$

$$\frac{\cancel{\sin^2\theta}}{\cos^2\theta} \cdot \frac{1}{\cancel{\sin^2\theta}} - \cot^2\theta = \sec^2\theta - \cot^2\theta$$

After we cancel out the $\sin^2\theta$, we're almost done:

$$\frac{\cancel{\sin^2\theta}}{\cos^2\theta}\cdot\frac{1}{\cancel{\sin^2\theta}}-\cot^2\theta=\sec^2\theta-\cot^2\theta$$

$$\frac{1}{\cos^2\theta}-\cot^2\theta=\sec^2\theta-\cot^2\theta$$

$$\sec^2\theta-\cot^2\theta=\sec^2\theta-\cot^2\theta$$

The trigonometric identities we have discussed in this section are summarized below:

Pythagorean Identities	Reciprocal Identities
$\sin^2\theta+\cos^2\theta=1$	$\tan\theta=\frac{\sin\theta}{\cos\theta}$
$\tan^2\theta+1=\sec^2\theta$	$\cot\theta=\frac{\cos\theta}{\sin\theta}$
$1+\cot^2\theta=\csc^2\theta$	$\sec\theta=\frac{1}{\cos\theta}$
	$\csc\theta=\frac{1}{\sin\theta}$

In the examples above and in the exercises, the form $\sin\theta$ or $\cos\theta$ is typically used, however any letter may be used to represent the angle in question so long as it is the SAME letter in all expressions. For example, we can say that:

$$\sin^2\theta+\cos^2\theta=1$$

or we can say that

$$\sin^2 x + \cos^2 x = 1$$

however:

$$\sin^2 \theta + \cos^2 x \neq 1$$

because θ and x could be different angles!

Exercises 3.1

In each problem verify the given trigonometric identity.

1. $\cos\theta(\sec\theta - \cos\theta) = \sin^2\theta$
2. $\tan\theta(\cot\theta + \tan\theta) = \sec^2\theta$
3. $\tan\theta(\csc\theta + \cot\theta) = \sec\theta + 1$
4. $\cot\theta(\sec\theta + \tan\theta) = \csc\theta + 1$
5. $\tan^2\theta\csc^2\theta - \tan^2\theta = 1$
6. $\sin^2\theta\cot^2\theta + \sin^2\theta = 1$
7. $\frac{\sin\theta\tan\theta+\sin\theta}{\tan\theta+\tan^2\theta} = \cos\theta$
8. $\frac{\cos\theta\cot\theta+\cos\theta}{\cot\theta+\cot^2\theta} = \sin\theta$
9. $\frac{(\sin\theta+\cos\theta)^2}{\cos\theta} = \sec\theta + 2\sin\theta$
10. $(\sin\theta + \cos\theta)^2 + (\sin\theta - \cos\theta)^2 = 2$
11. $\cos\theta(\tan\theta + \cot\theta) = \csc\theta$
12. $\sin\theta(\cot\theta + \tan\theta) = \sec\theta$
13. $\frac{\cos\theta}{\tan\theta} = \csc\theta - \sin\theta$
14. $\frac{\sin\theta}{\cot\theta} = \sec\theta - \cos\theta$
15. $\frac{\csc\theta}{\cos\theta} - \frac{\cos\theta}{\csc\theta} = \frac{\cot^2\theta+\sin^2\theta}{\cot\theta}$
16. $\frac{\sec\theta+\csc\theta}{\tan\theta+\cot\theta} = \sin\theta + \cos\theta$
17. $\frac{\sin\theta}{1+\sin\theta} - \frac{\sin\theta}{1-\sin\theta} = -2\tan^2\theta$
18. $\frac{\cos\theta}{1+\cos\theta} - \frac{\cos\theta}{1-\cos\theta} = -2\cot^2\theta$
19. $\frac{\cot\theta}{1+\csc\theta} - \frac{\cot\theta}{1-\csc\theta} = 2\sec\theta$
20. $\frac{\tan\theta}{1+\sec\theta} - \frac{\tan\theta}{1-\sec\theta} = 2\csc\theta$
21. $\frac{\sec^2\theta}{1+\cot^2\theta} = \tan^2\theta$
22. $\frac{\csc^2\theta}{1+\tan^2\theta} - \cot^2\theta$
23. $\sec^4\theta - \sec^2\theta = \tan^4\theta + \tan^2\theta$
24. $\csc^4\theta - \csc^2\theta = \cot^4\theta + \cot^2\theta$
25. $1 - \frac{\cos^2\theta}{1+\sin\theta} = \sin\theta$
26. $1 - \frac{\sin^2\theta}{1+\cos\theta} = \cos\theta$
27. $\frac{\sec\theta}{\csc\theta} + \frac{\sin\theta}{\cos\theta} = 2\tan\theta$
28. $\frac{1-\sin\theta}{\cos\theta} + \frac{\cos\theta}{1-\sin\theta} = 2\sec\theta$
29. $\frac{\cos\theta}{1+\sin\theta} + \frac{1+\sin\theta}{\cos\theta} = 2\sec\theta$
30. $\frac{\tan\theta-\cot\theta}{\tan\theta+\cot\theta} = \sin^2\theta - \cos^2\theta$
31. $\frac{\sec\theta-\cos\theta}{\sec\theta+\cos\theta} = \frac{\sin^2\theta}{1+\cos^2\theta}$
32. $\frac{\sec\theta+\tan\theta}{\cot\theta+\cos\theta} = \tan\theta\sec\theta$

3.2 Double-Angle Identities

In this section we will include several new identities to the collection we established in the previous section. These new identities are called "Double-Angle Identities" because they typically deal with relationships between trigonometric functions of a particular angle and functions of "two times" or double the original angle.

To establish the validity of these identities we need to use what are known as the Sum and Difference Identities. These are identities that deal with expressions such as $\sin(\alpha + \beta)$. First we will establish an expression that is equivalent to $\cos(\alpha - \beta)$.

Let's start with the unit circle:

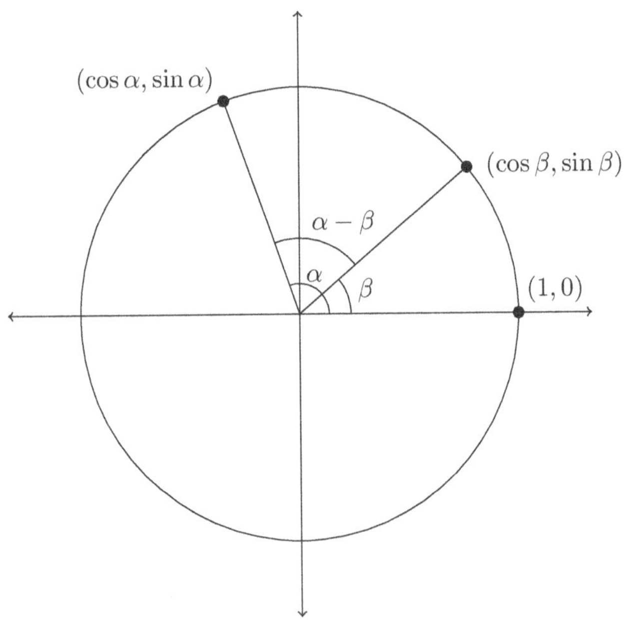

If we rotate everything in this picture clockwise so that the point labeled $(\cos\beta, \sin\beta)$ slides down to the point labeled $(1, 0)$, then the angle of rotation in the diagram will be $\alpha - \beta$ and the corresponding point on the edge of the circle will be: $(cos(\alpha - \beta), \sin(\alpha - \beta))$.

The diagram that represents this rotation is on the next page.

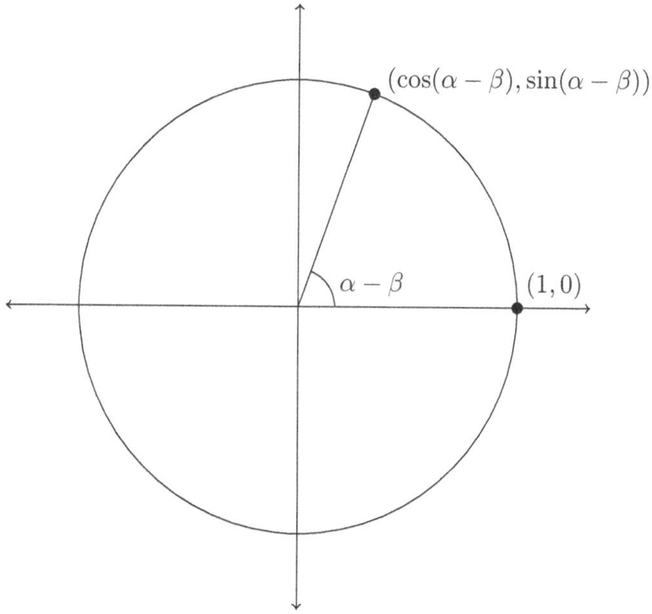

Since the the second diagram is created by rotating the lines and points from the first diagram, the distance between the points $(\cos\alpha, \sin\alpha)$ and $(\cos\beta, \sin\beta)$ in the first diagram is the same as the distance between $(\cos(\alpha-\beta), \sin(\alpha-\beta))$ and the point $(1, 0)$ in the second diagram.

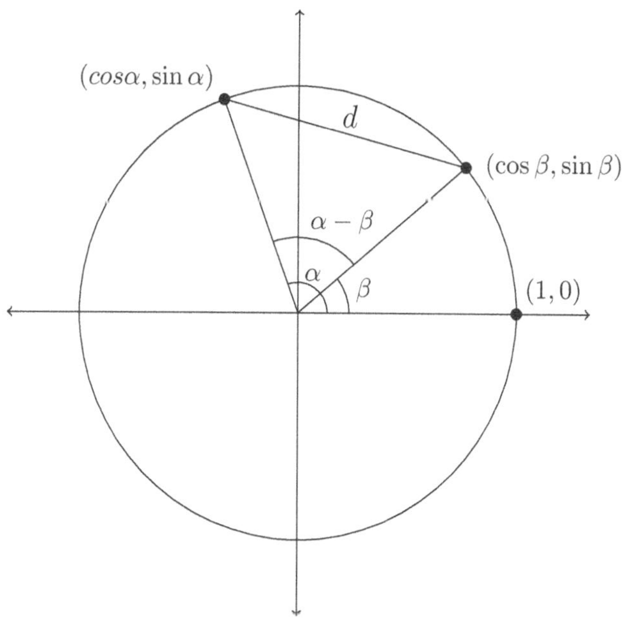

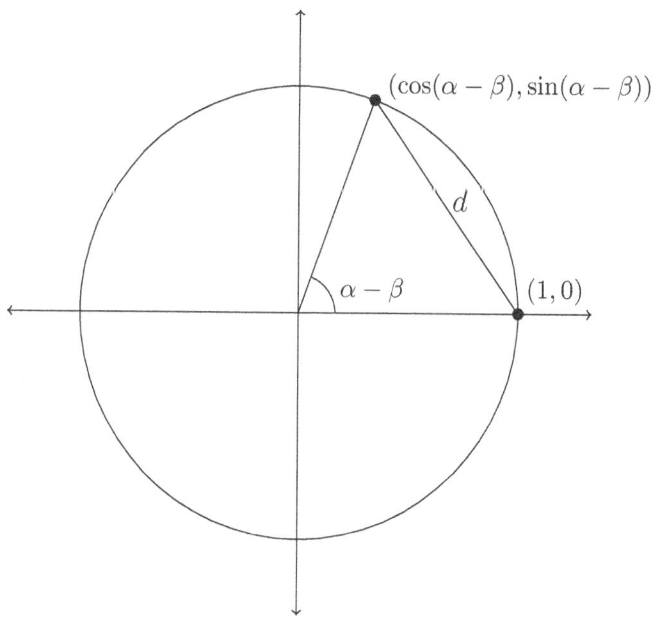

In the diagram above the length of d in each picture is the same.

We can represent this distance d with the distance formula used to calculate the distance between two points in the coordinate plane:

The distance between the points (x_1, y_1) and (x_2, y_2) is

$$d = \sqrt{(x_2 - x_1)^2 + (y_2 - y_1)^2}$$

So, in the first diagram the distance d will be:

$$d = \sqrt{(\cos\alpha - \cos\beta)^2 + (\sin\alpha - \sin\beta)^2}$$

In the second diagram the distance d will be:

$$d = \sqrt{(\cos(\alpha - \beta) - 1)^2 + (\sin(\alpha - \beta) - 0)^2}$$

Since these distances are the same, we can set them equal to each other:

$$\sqrt{(\cos\alpha - \cos\beta)^2 + (\sin\alpha - \sin\beta)^2} = \sqrt{(\cos(\alpha - \beta) - 1)^2 + (\sin(\alpha - \beta) - 0)^2}$$

We'll square both sides to clear the radicals:

$$(\cos\alpha - \cos\beta)^2 + (\sin\alpha - \sin\beta)^2 = (\cos(\alpha - \beta) - 1)^2 + (\sin(\alpha - \beta) - 0)^2$$

Next, we'll rewrite $(\sin(\alpha - \beta) - 0)^2$ as $\sin^2(\alpha - \beta)$:

$$(\cos\alpha - \cos\beta)^2 + (\sin\alpha - \sin\beta)^2 = (\cos(\alpha - \beta) - 1)^2 + \sin^2(\alpha - \beta)$$

Now we'll work to simplify the expressions on the left-hand side of this equation.

$$(\cos\alpha - \cos\beta)^2 + (\sin\alpha - \sin\beta)^2 = (\cos(\alpha-\beta) - 1)^2 + \sin^2(\alpha-\beta)$$

First, each one needs to be squared:

$$(\cos\alpha - \cos\beta)^2 = \cos^2\alpha - 2\cos\alpha\cos\beta + \cos^2\beta$$

$$(\sin\alpha - \sin\beta)^2 = \sin^2\alpha - 2\sin\alpha\sin\beta + \sin^2\beta$$

So, the left-hand side will now be:

$$\cos^2\alpha - 2\cos\alpha\cos\beta + \cos^2\beta + \sin^2\alpha - 2\sin\alpha\sin\beta + \sin^2\beta$$

If we rearrange this a little, it will simplify nicely:

$$\cos^2\alpha - 2\cos\alpha\cos\beta + \cos^2\beta + \sin^2\alpha - 2\sin\alpha\sin\beta + \sin^2\beta$$

$$\sin^2\alpha + \cos^2\alpha + \sin^2\beta + \cos^2\beta - 2\cos\alpha\cos\beta - 2\sin\alpha\sin\beta$$

Notice the Pythagorean Identities at the front of this expression - these are each equal to 1:

$$\sin^2\alpha + \cos^2\alpha + \sin^2\beta + \cos^2\beta - 2\cos\alpha\cos\beta - 2\sin\alpha\sin\beta$$

$$1 + 1 - 2\cos\alpha\cos\beta - 2\sin\alpha\sin\beta$$

$$2 - 2\cos\alpha\cos\beta - 2\sin\alpha\sin\beta$$

$$2(1 - \cos\alpha\cos\beta - \sin\alpha\sin\beta)$$

Now that we've simplified the left-hand side, we'll simplify the right-hand side. First we'll square the expression $(\cos(\alpha - \beta) - 1)^2$:

$$(\cos(\alpha - \beta) - 1)^2 = \cos^2(\alpha - \beta) - 2\cos(\alpha - \beta) + 1$$

So, the right-hand side is now:

$$\cos^2(\alpha - \beta) - 2\cos(\alpha - \beta) + 1 + \sin^2(\alpha - \beta)$$

If we rearrange this expression, we'll again have a nice Pythagorean Identity:

$$\sin^2(\alpha - \beta) + \cos^2(\alpha - \beta) - 2\cos(\alpha - \beta) + 1$$

$$1 - 2\cos(\alpha - \beta) + 1$$

$$2 - 2\cos(\alpha - \beta)$$

$$2(1 - \cos(\alpha - \beta))$$

So the left-hand side was equal to:

$$2(1 - \cos\alpha\cos\beta - \sin\alpha\sin\beta)$$

And the right-hand side was equal to:

$$2(1 - \cos(\alpha - \beta))$$

So, our original statement in simplified form is:

$$2(1 - \cos\alpha\cos\beta - \sin\alpha\sin\beta) = 2(1 - \cos(\alpha - \beta))$$

If we divide by 2 on both sides, we'll have:

$$1 - \cos\alpha\cos\beta - \sin\alpha\sin\beta = 1 - \cos(\alpha - \beta)$$

then subtract 1

$$-\cos\alpha\cos\beta - \sin\alpha\sin\beta = -\cos(\alpha - \beta)$$

and multiply through by -1

$$\cos\alpha\cos\beta + \sin\alpha\sin\beta = \cos(\alpha - \beta)$$

So, $\cos(\alpha - \beta) = \cos\alpha\cos\beta + \sin\alpha\sin\beta$.

This will help us to generate the double-angle formulas, but to do this, we don't want $\cos(\alpha - \beta)$, we want $\cos(\alpha + \beta)$ (you'll see why in a minute).

So, to change this around, we'll use identities for negative angles. Recall that in the fourth quadrant the sine function is negative and the cosine function is positive. For this reason, $\sin(-\theta) = -\sin(\theta)$ and $\cos(-\theta) = \cos(\theta)$.

Now we can say that $\cos(2\theta) = \cos(\theta + \theta) = \cos(\theta - (-\theta))$. Going back to our identity for $\cos(\alpha - \beta)$, we can say that:

$$\begin{aligned}
\cos(\theta - (-\theta)) &= \cos\theta\cos(-\theta) + \sin\theta\sin(-\theta)\\
\cos(\theta - (-\theta)) &= \cos\theta\cos\theta + \sin\theta(-\sin\theta)\\
\cos(\theta - (-\theta)) &= \cos\theta\cos\theta - \sin\theta\sin\theta\\
\cos(\theta - (-\theta)) &= \cos^2\theta - \sin^2\theta\\
\cos(\theta + \theta) &= \cos^2\theta - \sin^2\theta\\
\cos(2\theta) &= \cos^2\theta - \sin^2\theta
\end{aligned}$$

This is the double-angle identity for the cosine: $\cos(2\theta) = \cos^2\theta - \sin^2\theta$.

This identity actually appears in any one of three forms because the Pythagorean Identities can be applied to this to change its appearance:

$$\cos(2\theta) = \cos^2\theta - \sin^2\theta$$
$$\cos(2\theta) = 1 - \sin^2\theta - \sin^2\theta$$
$$\cos(2\theta) = 1 - 2\sin^2\theta$$

If we substitute for the $\sin^2\theta$ term:

$$\cos(2\theta) = \cos^2\theta - \sin^2\theta$$
$$\cos(2\theta) = \cos^2\theta - (1 - \cos^2\theta)$$
$$\cos(2\theta) = \cos^2\theta - 1 + \cos^2\theta$$
$$\cos(2\theta) = 2\cos^2\theta - 1$$

So, the three forms of the cosine double angle identity are:

$$\begin{aligned}\cos(2\theta) &= \cos^2\theta - \sin^2\theta \\ &= 2\cos^2\theta - 1 \\ &= 1 - 2\sin^2\theta\end{aligned}$$

The double-angle identity for the sine function uses what is known as the co-function identity. Remember that, in a right triangle, the sine of one angle is the same as the cosine of its complement (which is the other acute angle). This is because the adjacent side for one angle is the opposite side for the other angle. The denominator in both cases is the hypotenuse, so the cofunctions of complementary angles are equal.

In the diagram below, we can see this more clearly:

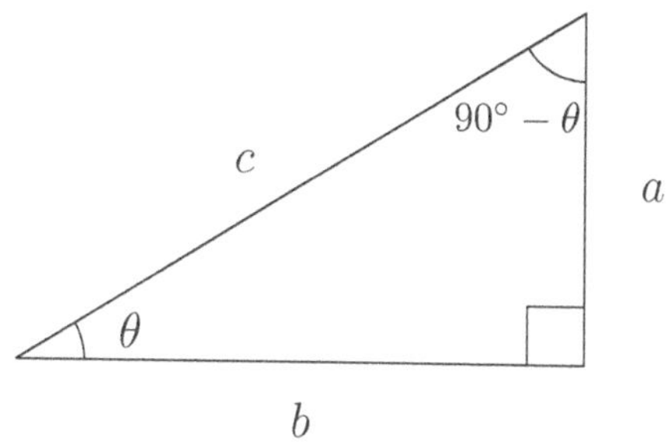

In the diagram above note that:

$$\sin\theta = \frac{a}{c} = \cos(90° - \theta)$$

So, if we want an identity for $\sin(\theta+\theta)$, we'll start with $\sin(\alpha+\beta)$ which is equivalent to $\cos(90° - (\alpha+\beta))$. We'll use a trick here and restate this as:

$$\begin{aligned}\sin(\alpha+\beta) &= \cos(90° - (\alpha+\beta)) \\ &= \cos(90° - \alpha - \beta) \\ &= \cos((90° - \alpha) - \beta) \\ &= \cos(90° - \alpha)\cos\beta + \sin(90° - \alpha)\sin\beta \\ &= \sin\alpha\cos\beta + \cos\alpha\sin\beta\end{aligned}$$

Now, we can use this to find an expression for $\sin 2\theta = \sin(\theta+\theta)$:

$$\begin{aligned}\sin 2\theta &= \sin(\theta+\theta) \\ &= \sin\theta\cos\theta + \cos\theta\sin\theta \\ &= 2\sin\theta\cos\theta\end{aligned}$$

Here is a summary of all the identities we've worked with in this chapter:

Pythagorean Identities

$\sin^2\theta + \cos^2\theta = 1$

$\tan^2\theta + 1 = \sec^2\theta$

$1 + \cot^2\theta = \csc^2\theta$

Reciprocal Identities

$\tan\theta = \frac{\sin\theta}{\cos\theta}$

$\cot\theta = \frac{\cos\theta}{\sin\theta}$

$\sec\theta = \frac{1}{\cos\theta}$

$\csc\theta = \frac{1}{\sin\theta}$

Double-Angle Identities

$\cos(2\theta) = \cos^2\theta - \sin^2\theta$

$\cos(2\theta) = 2\cos^2\theta - 1$

$\cos(2\theta) = 1 - 2\sin^2\theta$

$\sin(2\theta) = 2\sin\theta\cos\theta$

Working problems involving double-angle identities is very similar to the other identities we've worked with previously - you just have more identities to choose from!

Example

Verify the given identity: $\cos 2x = \frac{1-\tan^2 x}{1+\tan^2 x}$

We have three possible identities to choose from for the left-hand side, so we'll wait on that for a moment while we simplify the right-hand side.

$$\frac{1-\tan^2 x}{1+\tan^2 x} = \frac{1-\tan^2 x}{\sec^2 x}$$

$$= \frac{1}{\sec^2 x} - \frac{\tan^2 x}{\sec^2 x}$$

$$= \cos^2 x - \frac{\left(\frac{\sin^2 x}{\cos^2 x}\right)}{\left(\frac{1}{\cos^2 x}\right)}$$

$$= \cos^2 x - \frac{\sin^2 x}{\cos^2 x} \cdot \frac{\cos^2 x}{1}$$

$$= \cos^2 x - \frac{\sin^2 x}{\cancel{\cos^2 x}} \cdot \frac{\cancel{\cos^2 x}}{1}$$

$$= \cos^2 x - \sin^2 x$$

This is one of the identities for $\cos(2\theta)$ so we can stop and simply state $\cos(2x) = \cos^2 x - \sin^2 x$

Exercises 3.2

In each problem verify the given trigonometric identity.

1. $\frac{2\sin x\cos x}{\cos^2 x - \sin^2 x} = \tan(2x)$

2. $\sec(2x) = \frac{\sec^2 x}{2 - \sec^2 x}$

3. $\sin(2x)\csc x = 2\cos x$

4. $\frac{2\cos x}{\sin(2x)} = \csc x$

5. $\frac{\cos(2x)}{\sin x} + \sin x = \frac{\cot x}{\sec x}$

6. $\frac{\sin x + \sin(2x)}{\sec x + 2} = \sin x\cos x$

7. $(\sin x + \cos x)^2 = 1 + \sin(2x)$

8. $(\sin^2 x - 1)^2 = \sin^4 x + \cos(2x)$

9. $2\cos x - \frac{\cos(2x)}{\cos x} = \sec x$

10. $\frac{1 + \cos(2x)}{1 - \cos(2x)} = \cot^2 x$

11. $\frac{\cos(2x)}{\sin^2 x} = \cot^2 x - 1$

12. $\frac{\cos(2x)}{\sin^2 x} = \csc^2 x - 2$

13. $\frac{\cot x - \tan x}{\cot x + \tan x} = \cos 2x$

14. $\sin 2x = \frac{2(\tan x - \cot x)}{\tan^2 x - \cot^2 x}$

15. $\frac{2\cos 2x}{\sin 2x} = \cot x - \tan x$

16. $\tan 2x = \frac{2}{\cot x - \tan x}$

17. $\frac{\sin x}{1 + \cos x} + \frac{1 + \cos x}{\sin x} = 2\csc x$

18. $\tan x + \cot x = 2\csc(2x)$

19. $\cos(2x) = \frac{\cot^2 x - 1}{\cot^2 x + 1}$

20. $\sin(2x) = \frac{2\tan x}{1 + \tan^2 x}$

21. $\frac{2\sin^2 x}{\sin(2x)} + \cot x = \sec x\csc x$

22. $\sec^2 x\cos(2x) = \sec^2 x - 2\tan^2 x$

23. $\frac{\cos(2x)}{\sin x} + \sin x = \csc x - \sin x$

24. $\frac{2\tan x - \sin(2x)}{2\sin^2 x} = \tan x$

3.3 Trigonometric Equations

In the previous section on trigonometric identities we worked with equations that would be true for all values of a particular angle θ. These are sort of like the algebraic equations whose solution set is "all real numbers," like $2x + 10 = 2(x+1)+8$. In this section, we will solve trigonometric equations whose solution set involves only certain values for the angle in question. Because of the cyclical nature of the angles we're working with, there will often be an infinite number of solutions although not "all real numbers."

Example 1

Here's an example. Suppose that we consider the equation $\sin x = 0.5$. Whether we use technology, a table or reasoning to solve this equation, it's clear that one solution is 30°. However, remember from the beginning of Chapter 2 that the sine function is positive in Quadrant II. That means that a second quadrant angle with a reference angle of 30°also has a sine equal to 0.5. Recall the ASTC diagram from Chapter 2:

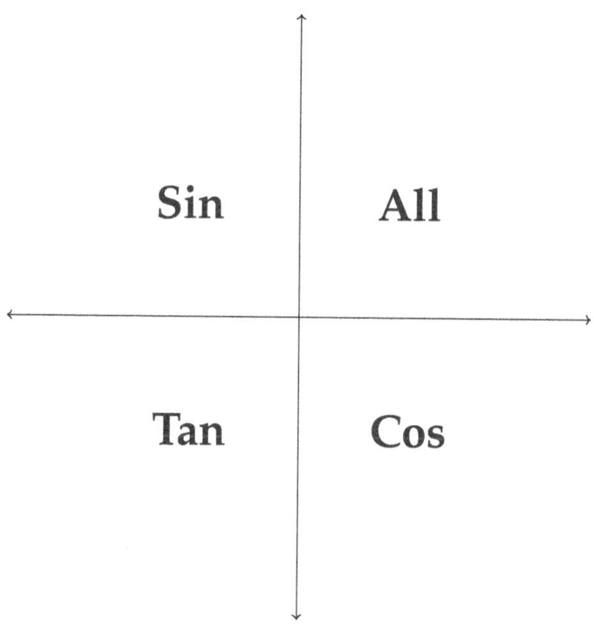

So, the sine function is positive in Quadrants I and II. This means that in addition to a solution of 30°, there is another solution in Quadrant II. As mentioned above, this second quadrant solution has a reference angle of 30°:

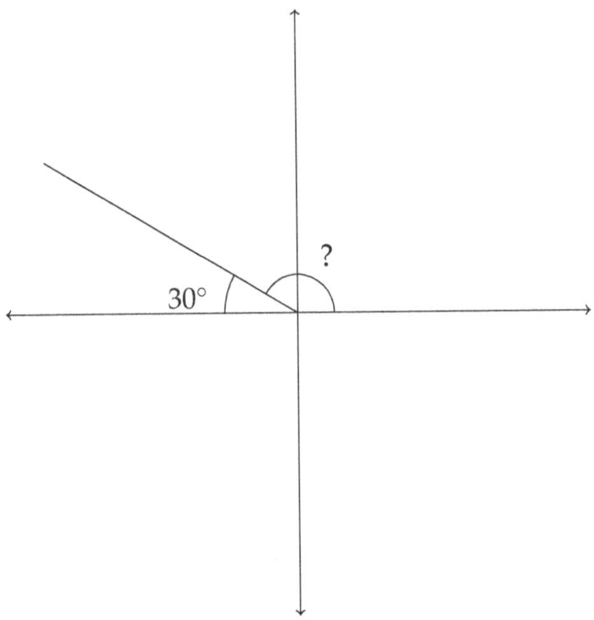

To find this angle, we simply subtract $180° - 30° = 150°$.

In Quadrant II, we subtract the reference angle from 180°.

In Quadrant III, we add the reference angle to 180°.

In Quadrant IV, we subtract the reference angle from 360°.

So, the solutions to the equation $\sin x = 0.5$ between 0°and 360°are $x = 30°, 150°$. In this chapter we will consider mainly solutions with this restriction:

$$0° \leq x < 360°.$$

The infinite solutions to this equation can be expressed as:

$$30° + n \cdot 360° \text{ and } 150° + n \cdot 360°.$$

Let's look at another example:

Example 2

Find all solutions of the given equation for $0^\circ \leq x < 360^\circ$.

$\tan x = 4$

Using a calculator to find $\tan^{-1}(4)$, we find that it returns an answer of $x \approx 75.96^\circ$. So this is the solution to the equation that lies in Quadrant I. The tangent function is also positive in Quadrant III, so we should also consider the third quadrant angle with a reference angle of 75.96°:

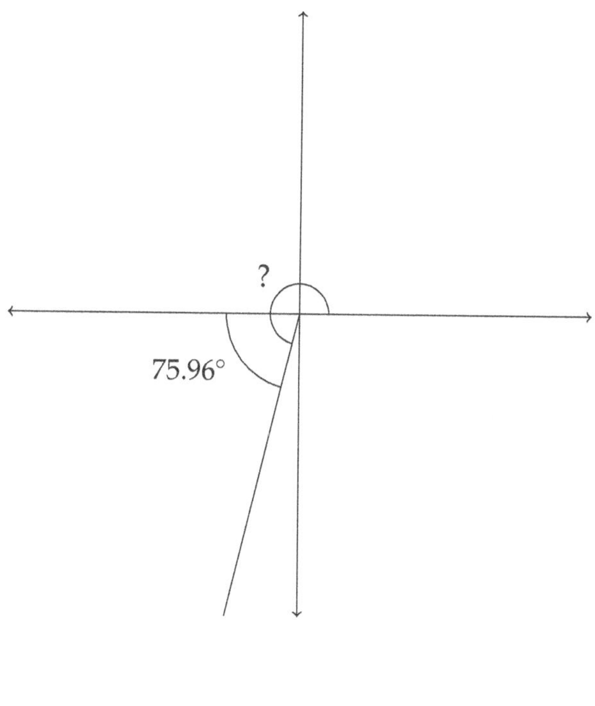

In Quadrant III, we add the reference angle to 180°:

$180^\circ + 75.96^\circ = 255.96^\circ$, so our solutions for this equation are $x \approx 75.96^\circ, 255.96^\circ$.

Often, calculators are programmed to return an angle value that is not between $0^\circ \leq x < 360^\circ$.

Example 3

Find all solutions of the given equation for $0° \leq x < 360°$.

$\sin x = -0.25$

Solving this on a TI calculator would generally return a value of $-14.5°$. However, $-14.5°$ is clearly not between 0°and 360°, so we need to use this information to find the solutions that are between 0°and 360°.

With the calculator returning a vlue of $-14.5°$, we know that the reference angle for all answers will be 14.5°. Knowing this, we can say that the sine is negative in Quadrants III and IV, so we'll need angles in those quadrants with reference angles of 14.5°.

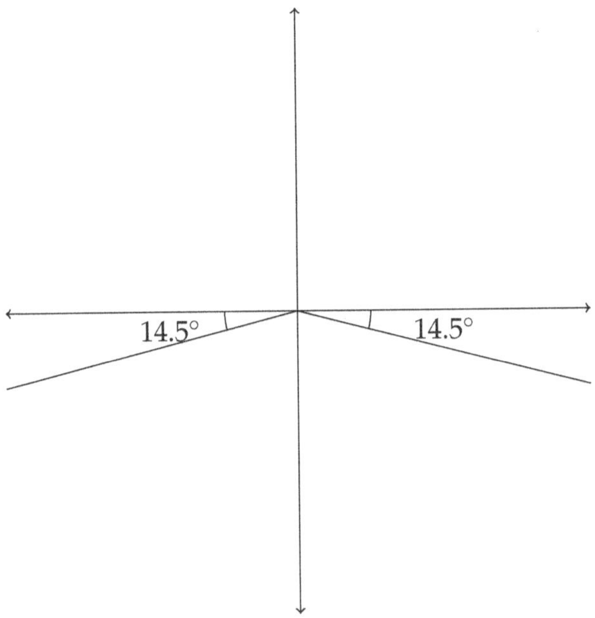

In Quadrant III we'll add 180°to the reference angle: $180° + 14.5° = 194.5°$

In Quadrant IV we'll subtract the reference angle from 360°: $360° - 14.5° = 345.5°$.

So, $x \approx 194.5°, 345.5°$.

Some trigonometric equations have no real number solutions. The equation $\sin x = 2$ has no real number solutions. Recall that the sine ratio was originally defined as the ratio of the side opposite an angle to the hypotenuse. The hypotenuse is always the longest side in a right triangle so there is no way the sine function could be greater than 1 if we're working with real-valued angles. However, in the same way that complex numbers are used to solve equations like $x^2 = -7$, complex-valued angles can be used to solve equations such as $\sin x = 2$. We won't go into this here, however, there is a relatively straightforward way to solve these equations.

If you encounter an equation like $\cos x = 3$ and are solving for values of x $0° \le x < 360°$, then the proper response is "no solution" or "no real solution." However, remember that the tangent function can take any value between $-\infty$ and ∞.

Example 4

Solving an equation that includes a reciprocal trigonometric function simply involves the extra step of finding the reciprocal:

Find all solutions of the given equation for $0° \le x < 360°$.

$\sec x = 12$

The trick here is to restate the equation so that we can use the preprogrammed values from a calculator to find the solution.

If $\sec x = 12$ then $\cos x = \frac{1}{12}$. Finding $\cos^{-1}(\frac{1}{12})$gives a solution of $x \approx 85.2°$.

The cosine and the secant are both positive in Quadrant IV, so we'll also want a fourth quadrant angle whose reference angle is 85.2°:

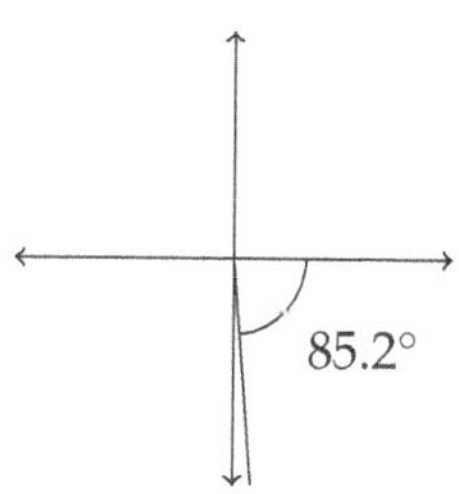

In Quadrant IV, we'll subtract the reference angle from 360°:

$$360° - 85.2° \approx 274.8°.$$

$$x \approx 85.2°, 274.8°$$

Example 5

Solving a quadratic trigonometric equation often involves the use of the quadratic formula:

Find all solutions of the given equation for $0° \leq x < 360°$:

$2\sin^2 x - \sin x - 2 = 0$

Using the quadratic formula we arrive at approximate values for $\sin x$ of $\sin x \approx -0.7808, 1.2808.$

The solution $\sin x \approx 1.2808$ yields no real solutions, so we will focus on solving $\sin x \approx -0.7808$

Finding $\sin^{-1}(-0.7808)$ gives us an answer of $\approx -51.3°$. This means our answers will lie in Quadrants III and IV with reference angles of 51.3°. In Quadrant III, we'll say $180° + 51.3° \approx 231.3°$. In Quadrant IV, we'll subtract the reference angle from 360°: $360° - 51.3° \approx 308.7°$.

So, $x \approx 231.3°, 308.7°$.

Exercises 3.3

Find all solutions for $0° \leq x < 360°$.
Round all angle measures to the nearest 10th of a degree.

1. $\cos x - 0.75 = 0$
2. $\sin x + 0.432 = 0$
3. $3\sin x - 5 = 0$
4. $\sin x - 4 = 0$
5. $3\sec x + 8 = 0$
6. $4\csc x + 9 = 0$
7. $3 - 5\sin x = 4\sin x + 1$
8. $4\cos x - 5 = \cos x - 3$
9. $3\tan^2 x + 2\tan x = 0$
10. $4\cos^2 x - \cos x = 0$
11. $3\cos^2 x + 5\cos x - 2 = 0$
12. $2\cot^2 x - 7\cot x + 3 = 0$
13. $2\tan^2 x - \tan x - 10 = 0$
14. $2\sin^2 x + 5\sin x + 3 = 0$
15. $2\cos^2 x - 5\cos x - 5 = 0$
16. $3\sin^2 x - \sin x - 1 = 0$

3.4 More Trigonometric Equations

When the solution to a trigonometric equation is one of the quadrantal angles (0°, 90°, 180°, 270°and so on), then determining all the solutions between 0°and 360°can work a little differently. The calculator will return some of these values, but in some cases it may not. If we go back to the unit circle, we can see this more clearly:

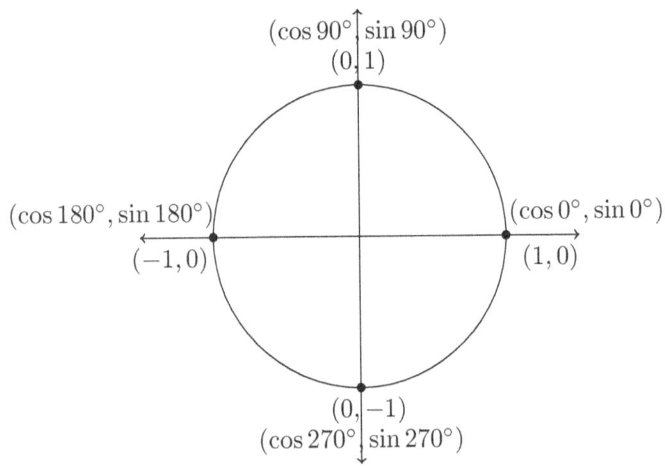

In the diagram above we can see the sine and cosine for $0°$, $90°$, $180°$, and $270°$. Since $\tan\theta = \frac{\sin\theta}{\cos\theta}$, then we can see that $\tan 0° = 0$, $\tan 90°$ is undefined, $\tan 180° = 0$ and $\tan 270°$ is also undefined.

The real issue with the quadrantal angles is finding $\sin^{-1}(0)$, $\cos^{-1}(0)$ or $\tan^{-1}(0)$. The calculator returns values of:

$$\sin^{-1}(0) = 0°$$

$$\cos^{-1}(0) = 90°$$

$$\tan^{-1}(0) = 0°$$

In each case, there is another possibility than differs from the given angle by 180°, so:

$$\sin^{-1}(0) = 0°, 180°$$

$$\cos^{-1}(0) = 90°, 270°$$

$$\tan^{-1}(0) = 0°, 180°$$

Let's look at how this is used in solving an equation:

Example 1

Solve the given equation for $0° \leq x < 360°$.

$\tan^2 x - \tan x = 0$

We could use the quadratic formula to solve this, but we can also solve by factoring:

$$\tan^2 x - \tan x = 0$$

$$\tan x(\tan x - 1) = 0$$

$$\tan x = 0 \text{ or } \tan x = 1$$

Using a calculator to find $\tan^{-1}(0)$ and $\tan^{-1}(1)$ returns values of $\tan^{-1}(0) = 0°$ and $\tan^{-1}(1) = 45°$. Once we know the reference angle for $\tan^{-1}(1)$, then we know that since the tangent is also positive in Quadrant III, the solutions here are 45°and 225°. The calculator returns an answer of 0°for $\tan^{-1}(0)$, but we just saw that $\tan 180° = 0$ as well.

The answers for this equation are $x = 45°, 225°, 0°, 180°$.

Another approach to solving trigonometric equations involves using Pythagorean Identities to make a substitution that so that the equation can be simply solved by the quadratic formula. Here's an example:

Example 2

Solve the given equation for $0° \le x < 360°$.

$\sin^2\theta - 6\cos\theta = 4$

Notice that, unlike the problems we saw in the previous section, this equation involves both the sine and the cosine. To remedy this, we can replace the $\sin^2\theta$ term with the expression $1 - \cos^2\theta$.

$$\sin^2\theta - 6\cos\theta = 4$$

$$1 - \cos^2\theta - 6\cos\theta = 4$$

$$0 = \cos^2\theta + 6\cos\theta + 3$$

using the quadratic formula:

$$\cos\theta \approx -5.449, -0.5505$$

Since $\cos^{-1}(-5.449)$ is not a real-valued angle, we can focus on the other answer: $\cos^{-1}(-0.5505) \approx 123.4°$. Since the cosine function is also negative in the third quadrant, we need to find the reference angle that will help us identify the third quadrant angle that is a solution for this equation:

$$180° - 123.4° = 56.6°$$

So the reference angle is $56.6°$.

$$180° + 56.6° = 236.6°$$

The solutions are $\theta \approx 123.4°, 236.6°$.

Example 3

Solve the given equation for $0° \leq x < 360°$.

$2\cos^2\theta - \sin\theta = \sin^2\theta + 1$

First, we'll substitute $1 - \sin^2\theta$ for the $\cos^2\theta$:

$$2\cos^2\theta - \sin\theta = \sin^2\theta + 1$$

$$2(1 - \sin^2\theta) - \sin\theta = \sin^2\theta + 1$$

$$2 - 2\sin^2\theta - \sin\theta = \sin^2\theta + 1$$

$$0 = 3\sin^2\theta + \sin\theta - 1$$

Solving this with the quadratic formula gives us solutions of $\sin\theta \approx -0.7676, 0.43426$.

$\sin^{-1}(-0.7676) \approx -50.1°$

$\sin^{-1}(0.43426) \approx 25.7°$

We'll work with the positive solution first. Since the sine is also positive in Quadrant II, the other angle will be $180° - 25.7° = 154.3°$.

For the negative solution, we know that the sine is negative in Quadrants III and IV, so with a reference angle of $50.1°$, in the third quadrant $180° + 50.1° = 230.1°$ and in the fourth quadrant $360° - 50.1° = 309.9°$.

The solution set is $\theta \approx 25.7°, 154.3°, 230.1°, 309.9°$.

Exercises 3.4

Solve the given equations for $0° \leq x < 360°$.

1. $9\sin^2\theta - 6\sin\theta = 1$
2. $4\cos^2\theta + 4\cos\theta = 1$
3. $\sec^2\alpha - 2\sec\alpha - 3 = 0$
4. $\csc^2\beta + 4\csc\beta - 10 = 0$
5. $\csc^2 x + 4\csc x - 7 = 0$
6. $3\cot^2 x - 3\cot x - 1 = 0$
7. $2\sin^2 x = 1 - \cos x$
8. $\cos^2\alpha + 4 = 2\sin\alpha - 3$
9. $\cos^2\beta - 3\sin\beta + 2\sin^2\beta = 0$
10. $\sin^2\theta = 2\cos\theta + 3\cos^2\theta$
11. $\sec^2 x = 2\tan x + 4$
12. $3\tan^2 x = \sec x + 2$
13. $\cos\alpha + 1 = 2\cos 2\alpha$
14. $\cos 2x - 3\sin x - 2 = 0$
15. $\csc^2\theta = \cot\theta + 5$
16. $\csc\theta + 5 = 2\cot^2\theta + 2$

Chapter 4

The Law of Sines
The Law of Cosines

In Chapter 1, we used the fundamental trigonometric relationships in right triangles to find unknown distances and angles. Unfortunately, in many problem-solving situations, it is inconvenient to use right triangle relationships. Therefore, from the right triangle relationships, we can derive relationships that can be used in any triangle.

4.1 The Law of Sines

The Law of Sines is based on right triangle relationships that can be created with the height of a triangle. Often, in this type of a problem, the angles are labeled with capital letters and their corresponding sides are labeled with lower case letters.

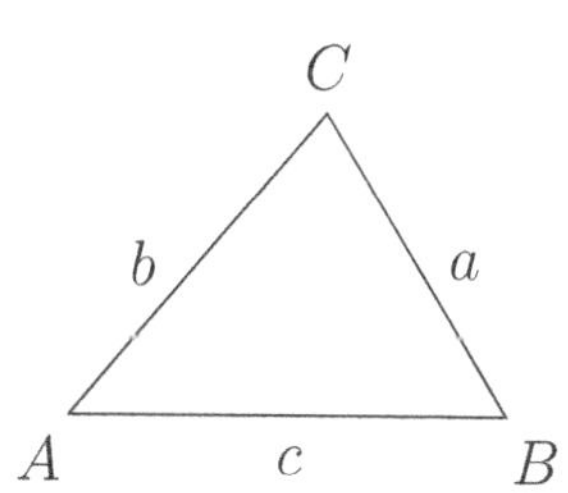

If we drop a perpendicular to the base of the triangle from the vertex point at $\angle C$, this creates two right triangles with which we can make use of the right triangle trigonometry covered in Chapter 1. This perpendicular would be the height of the triangle.

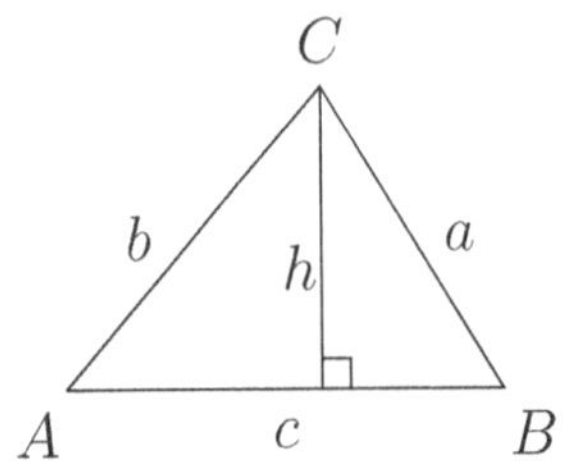

The Law of Sines is derived from this configuration and allows us to calculate the value of sides and angles in a triangle without a right angle, based on information about known sides and angles. Given the right triangles in the diagram above, we can see that:

$$\sin B = \frac{h}{a}$$

and

$$\sin A = \frac{h}{b}$$

Clearing the denominator in each fraction, we can see that:

$$a \sin B = h$$

and

$$b \sin A = h$$

so

$$a \sin B = b \sin A$$

To put this in the form in which the Law of Sines is normally stated, we can divide on both sides of the previous expression by ab:

$$a \sin B = b \sin A$$

$$\frac{a \sin B}{ab} = \frac{b \sin A}{ab}$$

$$\frac{\sin B}{b} = \frac{\sin A}{a}$$

A similar process will show that $\frac{\sin C}{c}$ is equivalent to $\frac{\sin B}{b}$ and $\frac{\sin A}{a}$. The diagram we derived this from used an acute triangle in which all the angles were less than 90°. The process to show that this is true for an obtuse triangle (which has one angle larger than 90°) is relatively simple and is left to the reader to discover or look up in another resource.

The Law of Sines

$$\frac{\sin A}{a} = \frac{\sin B}{b} = \frac{\sin C}{c}$$

Sometimes it is handy to set up a problem with the side lengths in the numerator:

The Law of Sines

$$\frac{a}{\sin A} = \frac{b}{\sin B} = \frac{c}{\sin C}$$

Example 1

Solve the triangle. Round side lengths to the nearest 100th.

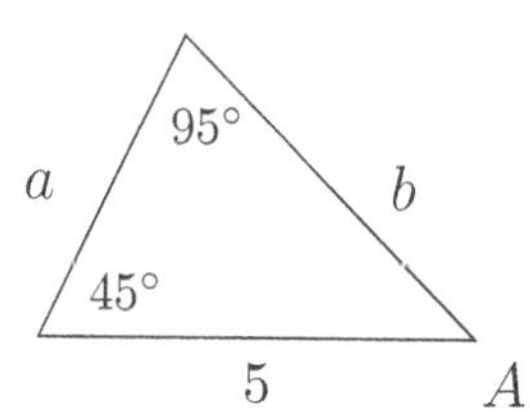

In this problem we're given two angles and one side. It's important that the side we're given corresponds to one of the known angles, otherwise we wouldn't be able to use the Law of Sines.

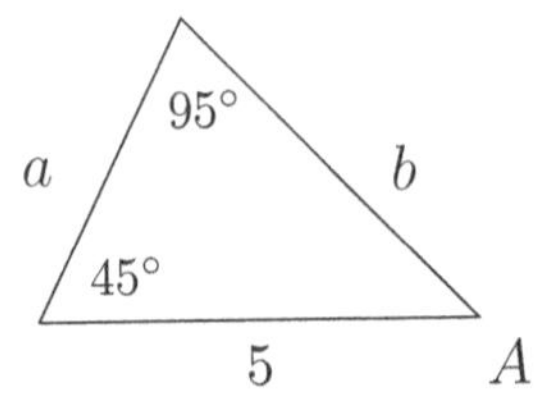

Since we know two of the angles, then the third will just be $180° - (45° + 95°) = 180° - 140° = 40° = \angle A$. To find the lengths of the unknown sides, we'll use the Law of Sines. We should start by choosing a side-angle pair for which we know both the side and the angle. In this case, we know that $\angle C = 95°$ and side $c = 5$.

$$\frac{c}{\sin C} = \frac{b}{\sin B}$$

$$\frac{5}{\sin 95°} = \frac{b}{\sin 45°}$$

If we multiply on both sides by $\sin 45°$, then

$$\sin 45° * \frac{5}{\sin 95°} = b$$

To arrive at an approximate value for $\sin 45° * \frac{5}{\sin 95°}$, we can say:

$$0.7071 * \frac{5}{0.9962} \approx b$$

$$3.55 \approx b$$

To find the length of side a, I would recommend that we use the exact side-angle pair that was given in the problem, rather than using the approximate value of side b that we just solved for.

This will make our value for side a more accurate:

$$\frac{c}{\sin C} = \frac{a}{\sin A}$$

$$\frac{5}{\sin 95^\circ} = \frac{a}{\sin 40^\circ}$$

Multiplying on both sides by $\sin 40^\circ$, then

$$\sin 40^\circ * \frac{5}{\sin 95^\circ} = a$$

To arrive at an approximate value for $\sin 40^\circ * \frac{5}{\sin 95^\circ}$, we can say:

$$0.6428 * \frac{5}{0.9962} \approx a$$

$$3.23 \approx a$$

$$\angle A = 40^\circ \qquad a \approx 3.23$$

$$\angle B = 45^\circ \qquad b \approx 3.55$$

$$\angle C = 95^\circ \qquad c = 5$$

Example 2

Some problems don't come with diagrams:

Solve the triangle if: $\angle A = 40^\circ$, $\angle B = 20^\circ$, $a = 2$.
Round side lengths to the nearest 100th.

Just as in the previous example, we can begin by finding the measure of the third angle $\angle C$. This would be $180^\circ - (40^\circ + 20^\circ) = 180^\circ - 60^\circ = 120^\circ = \angle C$

To find the missing sides, we should use the complete side-angle pair that is given in the problem: $\angle A = 40^\circ$ and $a = 2$.

We can find side b first or side c first, it doesn't matter which:

$$\frac{a}{\sin A} = \frac{b}{\sin B}$$

$$\frac{2}{\sin 40^\circ} = \frac{b}{\sin 20^\circ}$$

$$\sin 20^\circ * \frac{2}{\sin 40^\circ} = b$$

Then,

$$0.3420 * \frac{2}{0.6428} \approx b$$

$$1.06 \approx b$$

For side c:

$$\frac{a}{\sin A} = \frac{c}{\sin C}$$

$$\frac{2}{\sin 40^\circ} = \frac{c}{\sin 120^\circ}$$

$$\sin 120^\circ * \frac{2}{\sin 40^\circ} = c$$

Then,

$$0.8660 * \frac{2}{0.6428} \approx c$$

$$2.69 \approx c$$

$\angle A = 40^\circ$	$a = 2$
$\angle B = 20^\circ$	$b \approx 1.06$
$\angle C = 120^\circ$	$c \approx 2.69$

Exercises 4.1

In each problem below, solve the triangle. Round side lengths to the nearest 100th.

1. 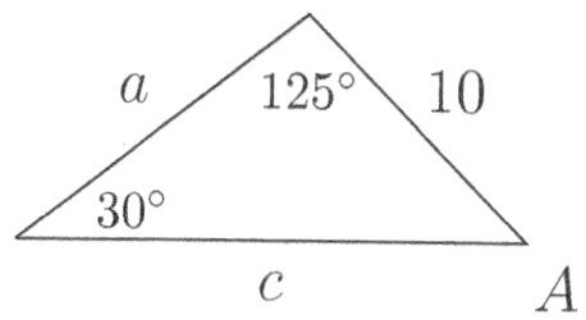

2. 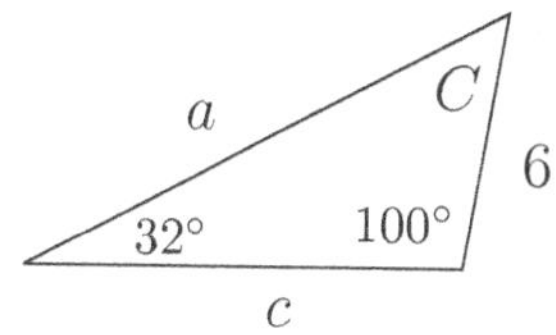

3. 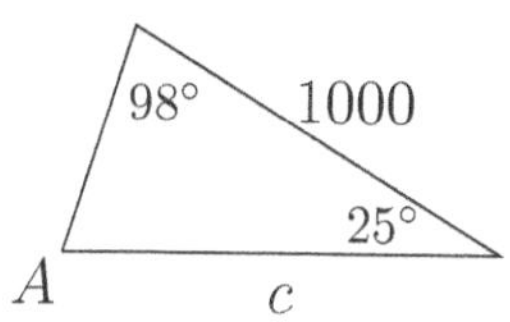

4. 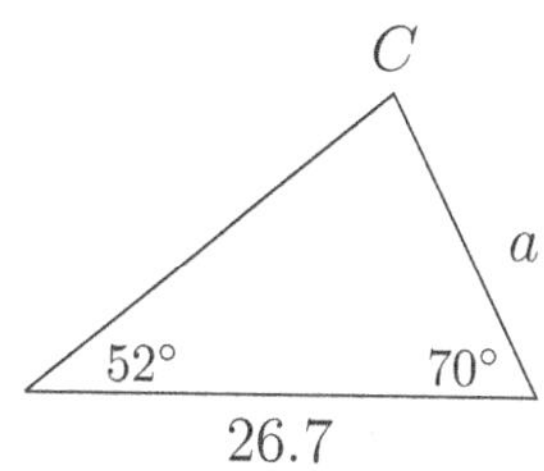

5. 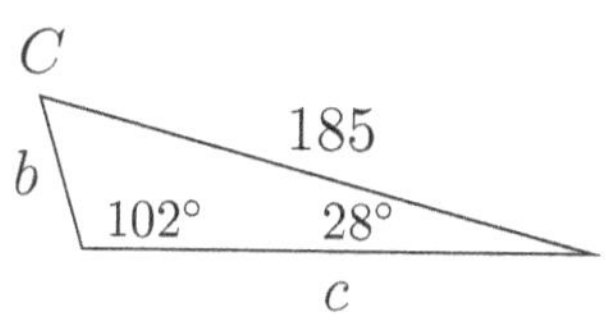

6. 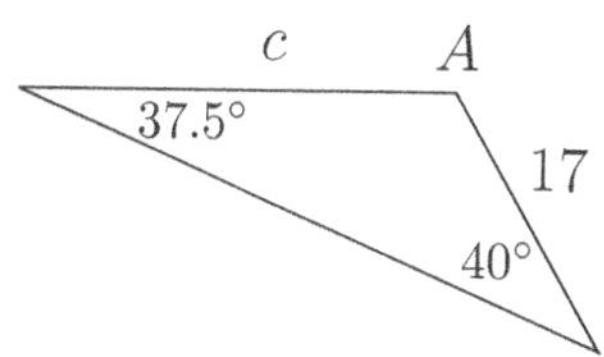

7. $\angle A = 50°, \quad \angle C = 27°, \quad a = 3$

8. $\angle B = 70°, \quad \angle C = 10°, \quad b = 5$

9. $\angle A = 110°, \quad \angle C = 30°, \quad c = 3$

10. $\angle A = 50°, \quad \angle B = 68°, \quad a = 230$

11. $\angle A = 23°, \quad \angle B = 110°, \quad c = 50$

12. $\angle A = 22°, \quad \angle B = 95°, \quad a = 420$

13. $\angle B = 10°, \quad \angle C = 100°, \quad c = 11$

14. $\angle A = 30°, \quad \angle C = 65°, \quad b = 10$

15. $\angle A = 82°, \quad \angle B = 65.4°, \quad b = 36.5$

16. $\angle B = 28°, \quad \angle C = 78°, \quad c = 44$

17. $\angle A = 42°, \quad \angle B = 61°, \quad a = 12$

18. $\angle A = 42.5°, \quad \angle B = 71.4°, \quad a = 215$

4.2 The Law of Sines: the ambiguous case

In all of the examples and problems in Section 4.1, notice that we were always given two angles and one side, although we could use the Law of Sines if we were given one angle and two sides (as long as one of the sides corresponded to the given angle). This is because when we use the Law of Sines to find an angle, an ambiguity can arise due to the sine function being positive in Quadrant I and Quadrant II.

We saw in Chapter 3 that multiple answers arise when we use the inverse trigonometric functions. For problems in which we use the Law of Sines given one angle and two sides, there may be one possible triangle, two possible triangles or no possible triangles. There are six different scenarios related to the ambiguous case of the Law of Sines: three result in one triangle, one results in two triangles and two result in no triangle.

One Triangle

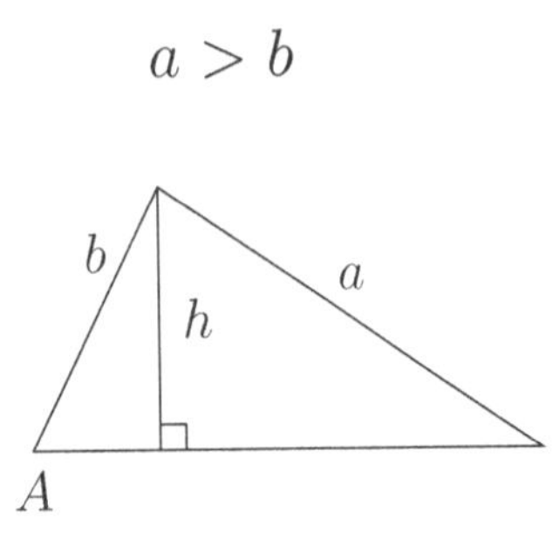

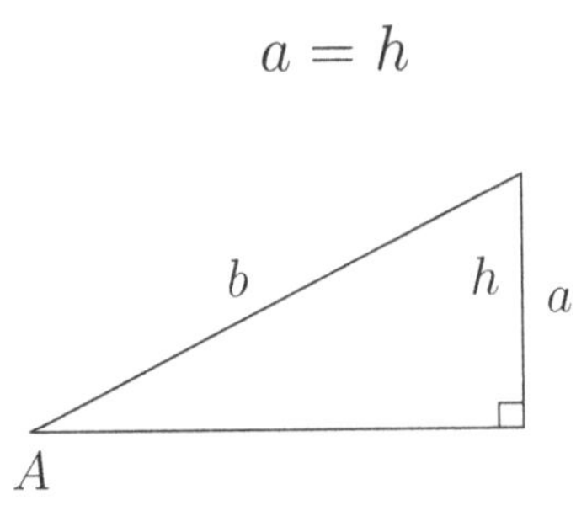

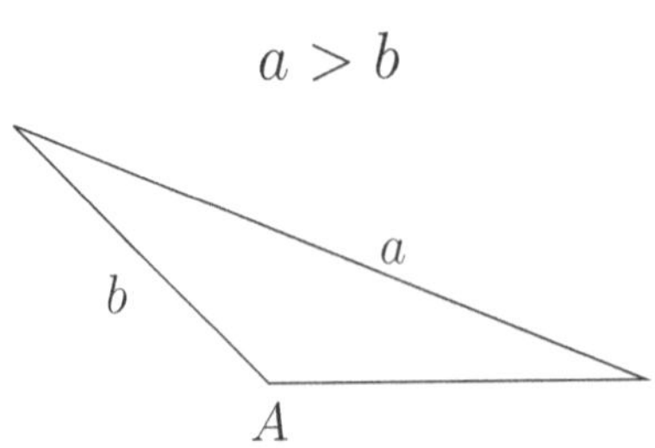

Two Triangles

$h < a < b$

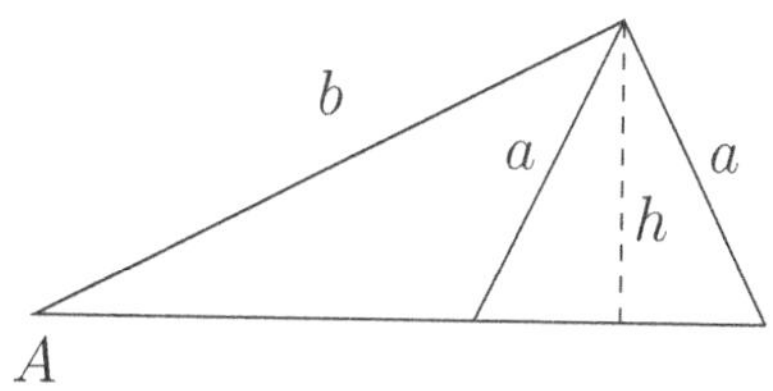

No Triangle

$a < h$

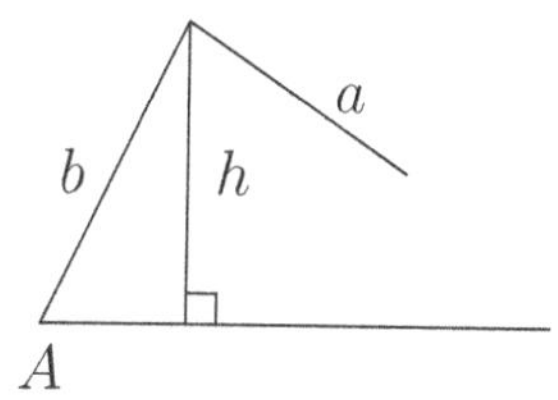

$a \leq b$

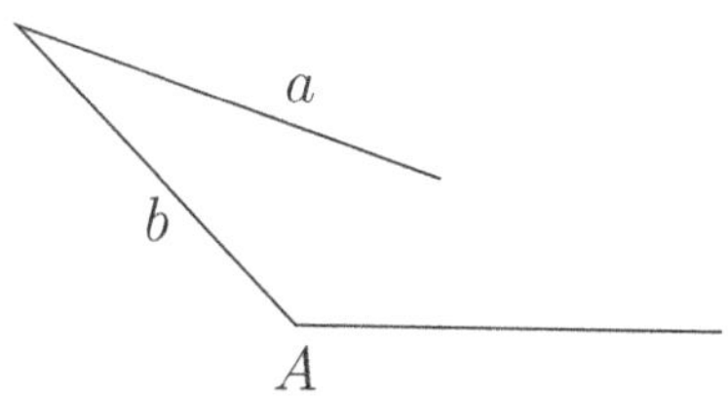

We'll look at three examples: one for one triangle, one for two triangles and one for no triangles.

Example 1

Solve the triangle if: $\angle A = 112°, \quad a = 45, \quad b = 24$

Round the angles and side lengths to the nearest 10th.

Using the Law of Sines, we can say that:

$$\frac{\sin 112°}{45} = \frac{\sin B}{24}$$

$$\frac{0.9272}{45} \approx \frac{\sin B}{24}$$

$$24 * \frac{0.9272}{45} \approx \sin B$$

$$0.4945 \approx \sin B$$

Then, we find $\sin^{-1}(0.4945) \approx 29.6°$. Remember from Chapter 3 that there is a Quadrant II angle that has $\sin\theta \approx 0.4945$, with a reference angle of 29.6°. So, $\angle B$ could also be $\approx 150.4°$. However, with $\angle A = 112°$, there is no way that another angle of 150.4°would fit inside the same triangle. For this reason, we know then that $\angle B$ must be 29.6°.

$$29.6° \approx B$$

So now

$$\angle A = 112°$$

$$\angle B \approx 29.6°$$

$$\text{and } \angle C = 180° - (112° + 29.6°) = 180° - 141.6° \approx 38.4°$$

$$\angle C \approx 38.4°$$

We already know that $a = 45$ and $b = 24$. To find side c, I would recommend using the most exact values possible in the Law of Sines calculation. This will provide the most accurate result in finding the length of side c.

$$\frac{45}{\sin 112°} = \frac{c}{\sin 38.4°}$$

$$\frac{45}{0.9272} \approx \frac{c}{0.6211}$$

$$0.6211 * \frac{45}{0.9272} \approx c$$

$$30.1 \approx c$$

$\angle A = 112°$	$a = 45$
$\angle B \approx 29.6°$	$b = 24$
$\angle C \approx 38.4°$	$c \approx 30.1$

Example 2

Solve the triangle if: $\angle A = 38°, \quad a = 40, \quad b = 52$

Round the angles and side lengths to the nearest 10th.

Using the Law of Sines, we can say that:

$$\frac{\sin 38°}{40} = \frac{\sin B}{52}$$

$$\frac{0.6157}{40} \approx \frac{\sin B}{52}$$

$$52 * \frac{0.6157}{40} \approx \sin B$$

$$0.8004 \approx \sin B$$

Just as in the previous example, we can find $\sin^{-1}(0.8004) \approx 53.2°$. But again, there is a Quadrant II angle whose sine has the same value ≈ 0.8004. The angle 126.8°has a sine ≈ 0.8004 and a reference angle of 53.2°. With $\angle A = 38°$, both of these angles (53.2°and 126.8°) could potentially fit in the triangle with angle A.

If we go back to the diagrams we looked at earlier in this section, we can see how this would happen:

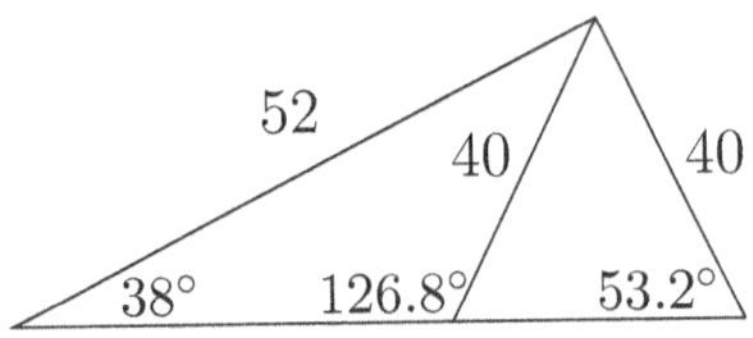

In the first possibility $\angle C$ would be $\approx 15.2°$.

In the second possibility $\angle C$ would be $\approx 88.8°$.

To find the two possible lengths for side c, we'll need to solve two Law of Sines calculations, one with $\angle C \approx 15.2°$ and one with the $\angle C \approx 88.8°$.

$$\frac{40}{\sin 38°} = \frac{c}{\sin 15.2°}$$

$$\frac{40}{0.6157} \approx \frac{c}{0.2622}$$

$$0.2622 * \frac{40}{0.6157} \approx c$$

$$17.0 \approx c$$

With $\angle C \approx 88.8°$:

$$\frac{40}{\sin 38°} = \frac{c}{\sin 88.8°}$$

$$\frac{40}{0.6157} \approx \frac{c}{0.9998}$$

$$0.9998 * \frac{40}{0.6157} \approx c$$

$$65.0 \approx c$$

So, our two possible solutions would be:

$$\angle A = 38° \qquad a = 40$$

$$\angle B \approx 126.8° \qquad b = 52$$

$$\angle C \approx 15.2° \qquad c \approx 17.0$$

OR

$$\angle A = 38° \qquad a = 40$$

$$\angle B \approx 53.2° \qquad b = 52$$

$$\angle C \approx 88.8° \qquad c \approx 65.0$$

Example 3

Solve the triangle if: $\angle B = 73°, \quad b = 51, \quad a = 92$

Round the angles and side lengths to the nearest 10th.

Using the Law of Sines, we can say that:

$$\frac{\sin 73°}{51} = \frac{\sin A}{92}$$

$$\frac{0.9563}{51} \approx \frac{\sin A}{92}$$

$$92 * \frac{0.9563}{51} \approx \sin A$$

$$1.7251 \approx \sin A$$

As we saw previously, no real-valued angle has a sine greater than 1. Therefore, no triangle is possible.

Exercises 4.2

In each problem, solve the triangle.
Round side lengths to the nearest 100th and angle measures to the nearest 10th.

1. $\angle A = 50°,\ b = 20,\ a = 32$
2. $\angle B = 40°,\ b = 4,\ c = 3$
3. $\angle A = 43°,\ a = 23,\ b = 29$
4. $\angle C = 20°,\ c = 43,\ a = 55$
5. $\angle B = 62°,\ b = 4,\ a = 5$
6. $\angle A = 75°,\ b = 8,\ \ a = 3$
7. $\angle B = 24°,\ a = 17,\ b = 8$
8. $\angle A = 40°,\ a = 4,\ c = 5$
9. $\angle A = 108°,\ a = 12,\ b = 7$
10. $\angle B = 117°,\ b = 19.6,\ c = 10.5$
11. $\angle A = 42°,\ a = 18,\ c = 11$
12. $\angle C = 27°,\ a = 42,\ \ c = 37$
13. $\angle C = 125°,\ c = 2.7,\ b = 5.2$
14. $\angle B = 115°,\ b = 68,\ \ a = 92$
15. $\angle A = 43°,\ a = 31,\ b = 37$
16. $\angle A = 28°,\ b = 3.5,\ \ a = 4.3$
17. $\angle C = 132°,\ c = 22,\ b = 16$
18. $\angle B = 114.2°,\ b = 87.2,\ \ a = 12.1$
19. $\angle B = 52°,\ c = 82.7,\ b = 70$
20. $\angle C = 65°,\ b = 7.6,\ \ c = 7.1$

4.3 The Law of Cosines

There are situations in which the Law of Sines cannot be used to solve a triangle. In the diagram below, we have information about two sides and the included angle:

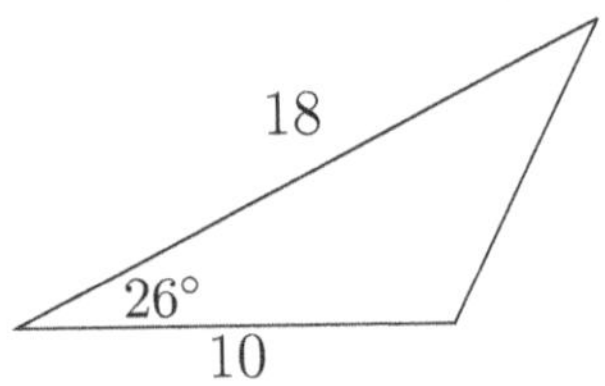

The problem above lacks a complete angle-side pair which is necessary to set up the Law of Sines calculation.

Another common situation involves a triangle in which all three sides are known but no angles are known:

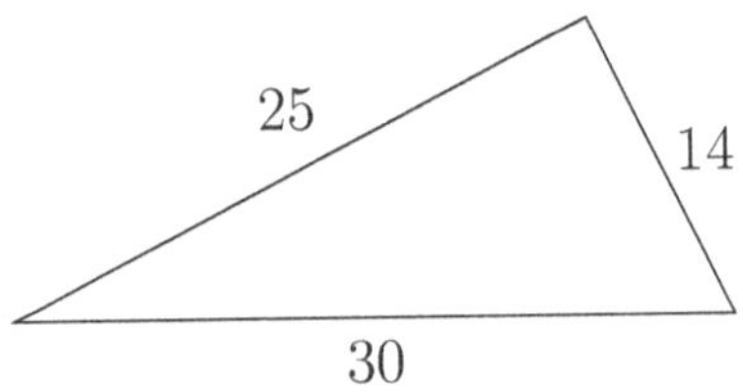

Again, the lack of an angle-side pair would prevent us from setting up a Law of Sines calculation.

The Law of Cosines is one way to get around this difficulty. Using the Law of Cosines is more complicated than using the Law of Sines, however, as we have just seen, the Law of Sines will not always be enough to solve a triangle.

To derive The Law of Cosines, we begin with an arbitrary triangle, like the one seen on the next page:

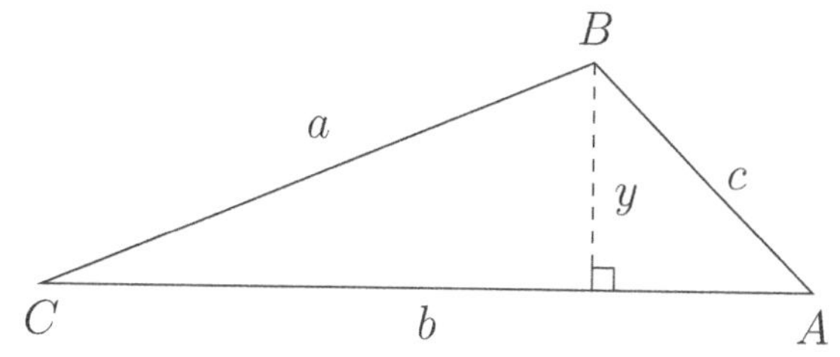

In this diagram we have taken the arbitrary triangle and created a perpendicular with length y. From this, we can say that $\sin C = \frac{y}{a}$ and that $a \sin C = y$.

We can split side AC into two pieces AD and CD, as seen below, and label the distance CD as x.

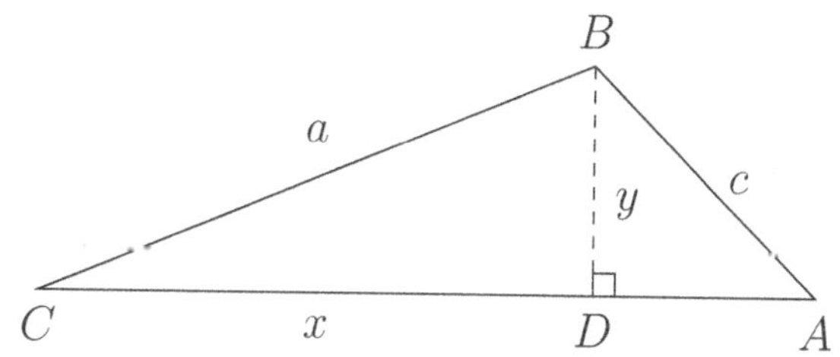

Then, we can say that $\cos C = \frac{x}{a}$ and that $a \cos C = x$.

If we then put this triangle onto the coordinate axes with $\angle C$ at the origin $(0, 0)$, we can derive the Law of Cosines. The coordinate of the vertex at $\angle B$ will be $(a \cos C, a \sin C)$, and the coordinates of the vertex at $\angle A$ will be $(b, 0)$.

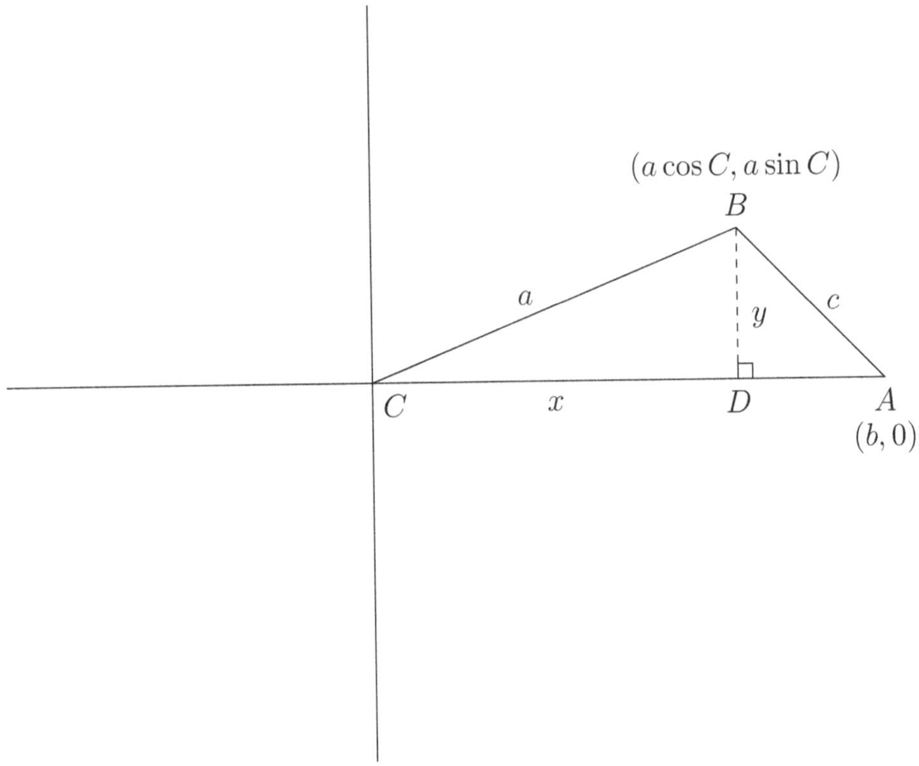

Using the distance formula, we can say that:

$$c = \sqrt{(a\cos C - b)^2 + (a\sin C - 0)^2}$$

Squaring both sides:

$$c^2 = (a\cos C - b)^2 + (a\sin C - 0)^2$$

and

$$c^2 = a^2\cos^2 C - 2ab\cos C + b^2 + a^2\sin^2 C$$

or

$$c^2 = a^2\sin^2 C + a^2\cos^2 C + b^2 - 2ab\cos C$$

Factoring out the a^2 and replacing $\sin^2 C + \cos^2 C$ with 1, we come out with one of the most common forms of the Law of Cosines:

$$c^2 = a^2 \sin^2 C + a^2 \cos^2 C + b^2 - 2ab \cos C$$

$$c^2 = a^2(\sin^2 C + \cos^2 C) + b^2 - 2ab \cos C$$

$$c^2 = a^2(1) + b^2 - 2ab \cos C$$

$$c^2 = a^2 + b^2 - 2ab \cos C$$

Any letter may be used to represent each of the sides, but the order that the letters are used in is very important. The side of the triangle isolated on the left hand side must correspond to the angle used on the right hand side.

The Law of Cosines

$$a^2 = b^2 + c^2 - 2bc \cos A$$

$$b^2 = a^2 + c^2 - 2ac \cos B$$

$$c^2 = a^2 + b^2 - 2ab \cos C$$

We'll look at three examples - two in which two sides and the included angle are given and one in which the three sides of the triangle are given.

Example 1

Solve the triangle: $\angle A = 38°,\ c = 17,\ b = 8$
Round angle measures and side lengths to the nearest 10th.

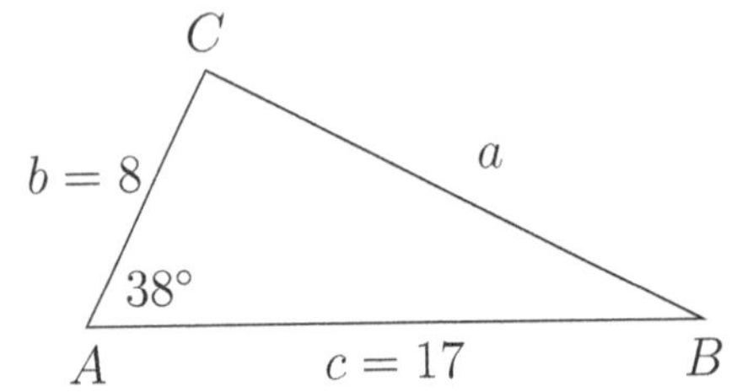

It's usually a good idea to see if you can use the Law of Sines first, since it is easier to calculate. In this case we can't because we don't have a complete angle-side pair. So, using the Law of Cosines to find side a:

$$a^2 = b^2 + c^2 - 2bc\cos A$$

$$a^2 = 8^2 + 17^2 - 2 * 8 * 17 * \cos 38°$$

$$a^2 \approx 64 + 289 - 272 * 0.7880$$

$$a^2 \approx 353 - 214.336$$

$$a^2 \approx 138.664$$

$$a \approx 11.8$$

Once we know that $a \approx 11.8$ we can use this to find the other angles using the Law of Sines. Because of the issue of the ambiguous case in using the Law of Sines, it's often a good idea to find the angles that correspond to the two shortest sides in the triangle, because if there is an obtuse angle in the triangle it will have to correspond to the longest side. If we find the two smaller angles, we can be assured that they will both be acute and we can subtract from 180°to find the largest angle.

$$\frac{\sin 38°}{11.8} = \frac{\sin B}{8}$$

$$8 * \frac{0.61566}{11.8} = \sin B$$

$$0.4174 \approx \sin B$$

$$24.7° \approx B$$

So with $\angle A = 38°$ and $\angle B \approx 24.7°$, then:

$$\angle C \approx 180° - (38° + 24.7°) \approx 180° - 62.7° \approx 117.3°$$

So, the angles and sides of the triangle would be:

$\angle A = 38°$	$a \approx 11.8$
$\angle B \approx 24.7°$	$b = 8$
$\angle C \approx 117.3°$	$c = 17$

In example 2, we'll look at a problem in which an obtuse angle is given.

Example 2

Solve the triangle: $\angle A = 110°,\ c = 30,\ b = 35$
Round angle measures and side lengths to the nearest 10th.

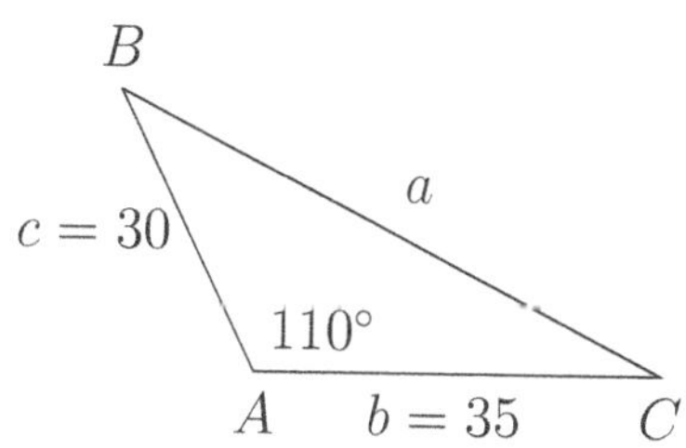

The calculation for this problem is slightly different from the last one because the cosine of 110°will be negative:

$$a^2 = b^2 + c^2 - 2bc\cos A$$

$$a^2 = 35^2 + 30^2 - 2 * 35 * 30 * \cos 110°$$

$$a^2 \approx 1225 + 900 - 2100 * (-0.3420)$$

$$a^2 \approx 2125 + 718.2$$

$$a^2 \approx 2843.2$$

$$a \approx 53.3$$

In this problem, since we were given an obtuse angle, then the other two angles must be acute and we don't have to worry about the ambiguous case in using the Law of Sines.

$$\frac{\sin 110°}{53.3} = \frac{\sin B}{35} \qquad \frac{\sin 110°}{53.3} = \frac{\sin C}{30}$$

$$35 * \frac{0.9397}{53.3} = \sin B \qquad 30 * \frac{0.9397}{53.3} = \sin C$$

$$0.61706 \approx \sin B \qquad 0.5289 \approx \sin C$$

$$38.1° \approx B \qquad 31.9° \approx C$$

So the angles and sides of the triangle would be:

$$\angle A = 110° \qquad a \approx 53.3$$

$$\angle B \approx 38.1° \qquad b = 35$$

$$\angle C \approx 31.9° \qquad c = 30$$

In example 3, we'll look at a problem in which three side lengths are given and we find an angle using the Law of Cosines.

Example 3

Solve the triangle: $a = 20, \; c = 9, \; b = 15$
Round angle measures to the nearest 10th.

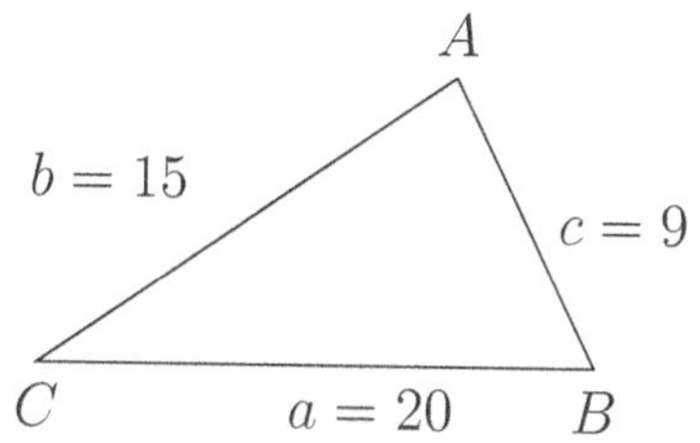

It doesn't matter which angle we choose to solve, but whichever angle we choose must correspond to the side isolated on the left-hand side of the formula. If we want to solve for $\angle B$, we would say:

$$b^2 = a^2 + c^2 - 2ac\cos B$$

$$15^2 = 20^2 + 9^2 - 2 * 20 * 9 * \cos B$$

$$225 = 400 + 81 - 360 * \cos B$$

$$225 = 481 - 360\cos B$$

$$-256 = -360\cos B$$

$$\frac{-256}{-360} = \frac{-360\cos B}{-360}$$

$$0.7\overline{1} = \cos B$$

$$0.7\bar{1} = \cos B$$

$$44.7^\circ \approx B$$

Once we know the measure of $\angle B$, we'll use this to find the measure of $\angle C$, which corresponds to side c, the smallest side. Then we'll subtract to find the biggest angle.

$$\frac{\sin 44.7^\circ}{15} = \frac{\sin C}{9}$$

$$9 * \frac{0.7034}{15} = \sin C$$

$$0.42204 \approx \sin C$$

$$25.0^\circ \approx C$$

So, with $\angle B \approx 44.7^\circ$ and $\angle C \approx 25.0^\circ$, then:

$$\angle A \approx 180^\circ - (44.7^\circ + 25.0^\circ) \approx 180^\circ - 69.7^\circ \approx 110.3^\circ$$

So the angles and sides of the triangle would be:

$\angle A = 110.3^\circ$	$a \approx 53.3$
$\angle B \approx 44.7^\circ$	$b = 35$
$\angle C \approx 25.0^\circ$	$c = 30$

If we had used the Law of Sines to find $\angle A$, the calculator would have returned the value of the reference angle for $\angle A$, rather than the angle that is actually in the triangle described in the problem!

Exercises 4.3

In each problem, solve the triangle.
Round side lengths to the nearest 100th and angle measures to the nearest 10th.

1. 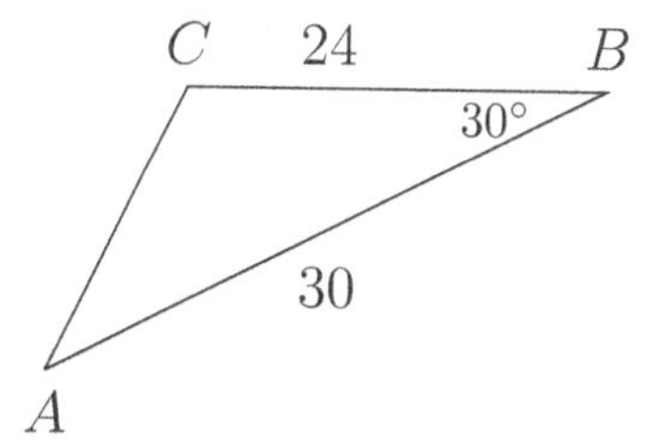

2. 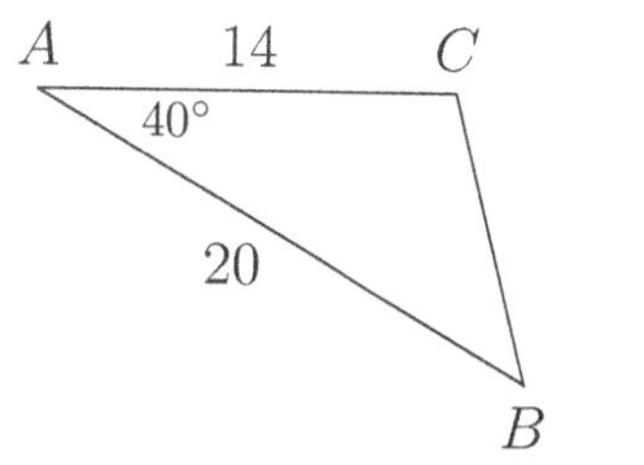

3. 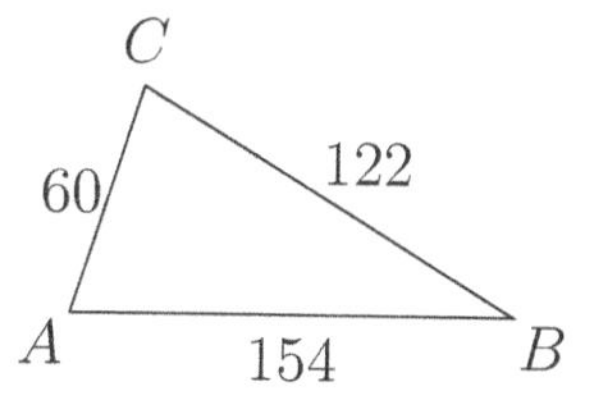

4. 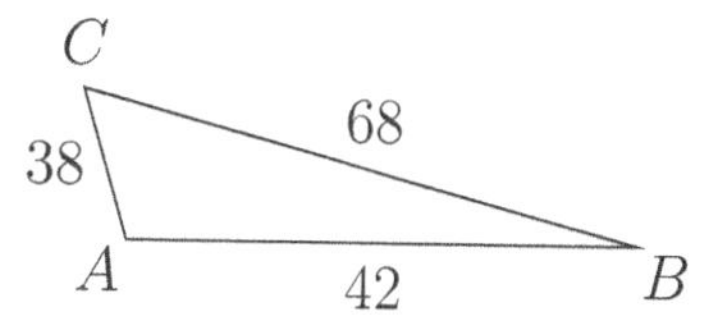

5. $\angle A = 52°, \ c = 27, \ b = 36$

6. $\angle B = 75°, \ a = 32, \ c = 59$

7. $\angle B = 135°, \ a = 12, \ c = 18$

8. $\angle C = 120°, \ b = 22, \ a = 30$

9. $a = 21, \ b = 26, \ c = 23$

10. $a = 11, \ b = 13, \ c = 17$

11. $a = 25, \ b = 32, \ c = 40$

12. $a = 60, \ b = 88, \ c = 120$

13. $\angle A = 77.4°, \ b = 444, \ c = 390$

14. $\angle B = 10°, \ a = 18, \ c = 30$

15. $a = 112.7, \ b = 96.5, \ c = 130.2$

16. $a = 4.7, \ b = 3.2, \ c = 5.9$

4.4 Applications

In the previous sections on applications, we saw situations in which right triangle trigonometry was used to find distances and angles. In this section, we will use the Law of Sines and the Law of Cosines to find distances and angles.

Example 1

A car travels along a straight road, heading west for 1 hour, then traveling on another straight road northwest for a half hour. If the speed of the car was a constant 50 mph how far is the car from its starting point?

First, let's draw a diagram:

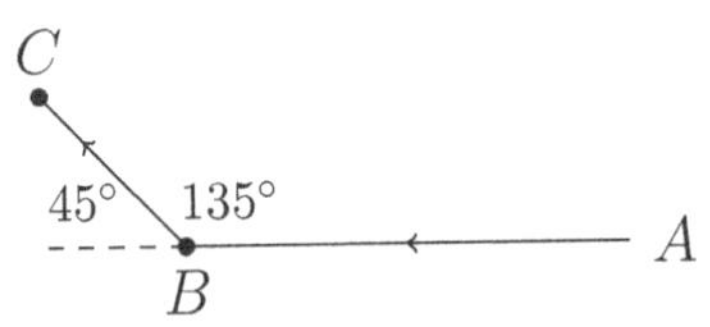

In the picture above, we know the angles $45°$ and $135°$ because of the direction the car was traveling. The direction northwest cuts exactly halfway between north and west creating a $45°$ angle. On the other side of this $45°$ angle is a $135°$ angle which is in the triangle we'll use to answer the question (triangle ABC).

The length of $\overline{AB}$ is 50 miles and the length of $\overline{BC}$ is 25 miles. This comes from the information about the speed and traveling time given in the problem. So the triangle we need to answer the question is pictured below:

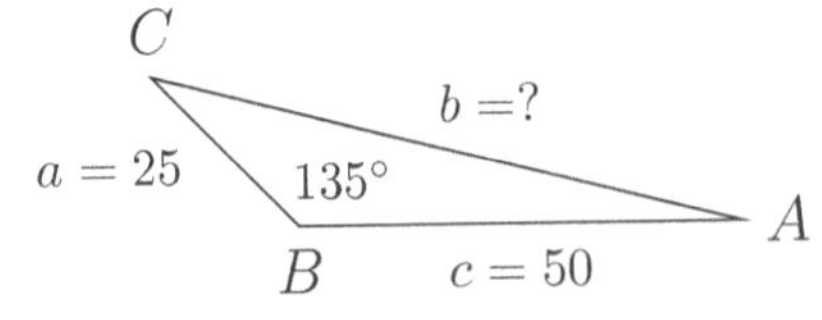

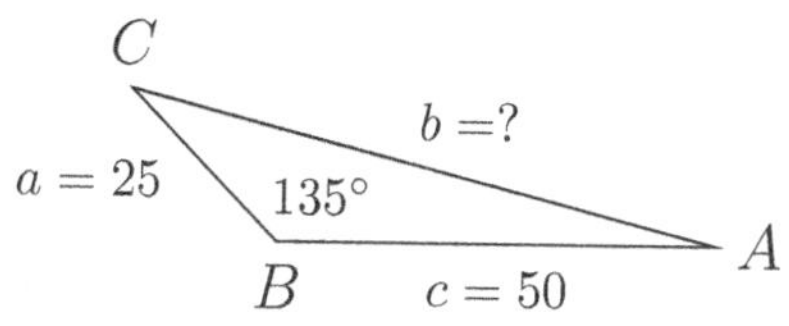

We can use the Law of Cosines to solve this problem:

$$b^2 = a^2 + c^2 - 2ac\cos B$$

$$b^2 = 25^2 + 50^2 - 2 * 25 * 50 * \cos 135°$$

$$b^2 \approx 625 + 2500 - 2500 * (-0.7071)$$

$$b^2 \approx 3125 + 1767.75$$

$$b^2 \approx 4892.75$$

$$b \approx 69.9 \text{ miles}$$

Example 2

A pilot flies an airplane in a straight path for 2.5 hours and then makes a course correction, heading 10°to the left of the original course. The pilot then flies in this direction for 1 hour. If the speed of the plane is a constant 350 mph, how far is the plane from its starting position?

Again, we'll start by making a diagram:

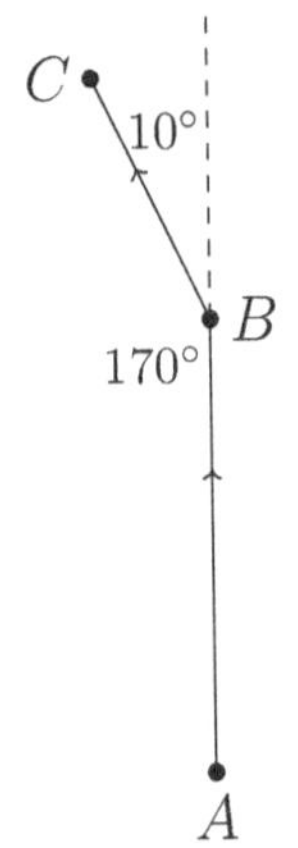

In this problem, we'll be working with triangle ABC, shown below. We can calculate the lengths of $\overline{AB}$ and $\overline{BC}$ from the information given in the problem and use this to calculate the length of $\overline{AC}$:

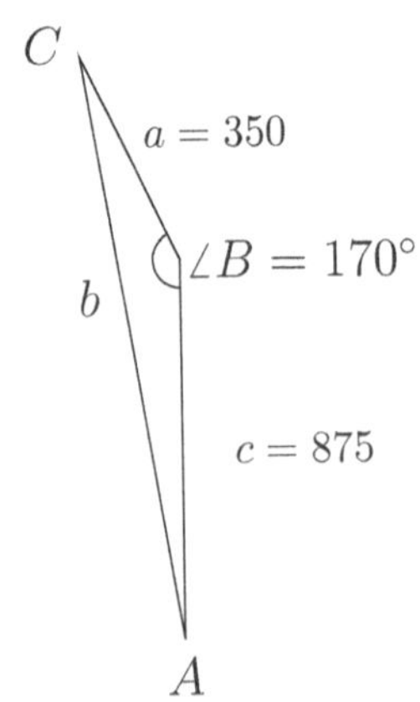

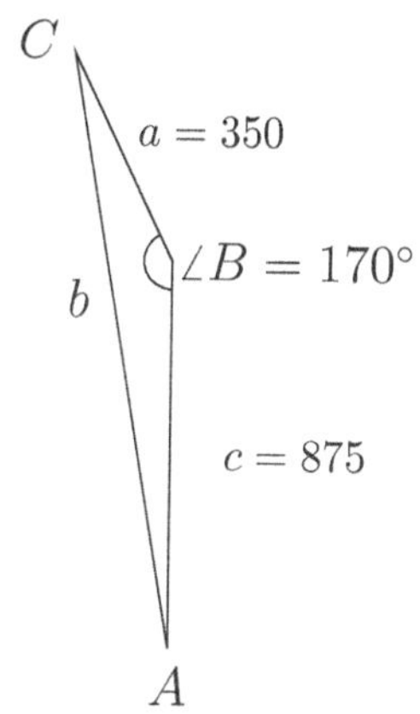

Using the Law of Cosines:

$$b^2 = a^2 + c^2 - 2ac\cos B$$

$$b^2 = 350^2 + 875^2 - 2 * 350 * 1050 * \cos 170°$$

$$b^2 \approx 122,500 + 765,625 - 735,000 * (-0.9848)$$

$$b^2 \approx 888,125 + 723,828$$

$$b^2 \approx 1,611,953$$

$$b \approx 1270 \text{ miles}$$

Example 3

A pilot leaves the airport in Bend, headed towards Corvallis with the bearing $N70°W$. He travels the 103 miles and makes a delivery before taking off and flying at a bearing of $N25°E$ for 72 miles to arrive in Portland.

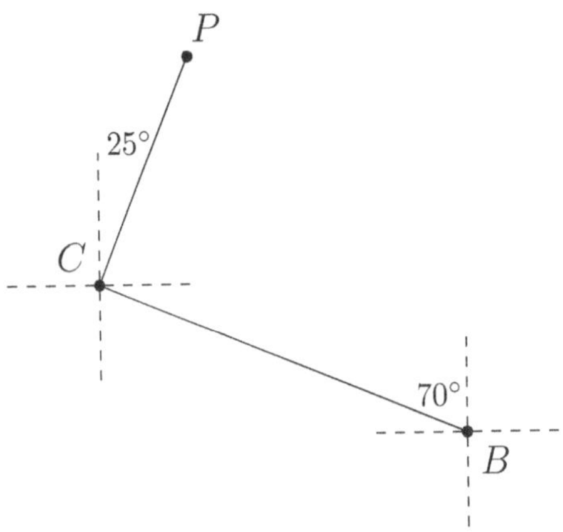

a) Based on this information, find the air distance between Portland and Bend.

b) Find the bearing from Portland to Bend.

In this problem, a diagram has been given. We'll amend this to make it into a triangle:

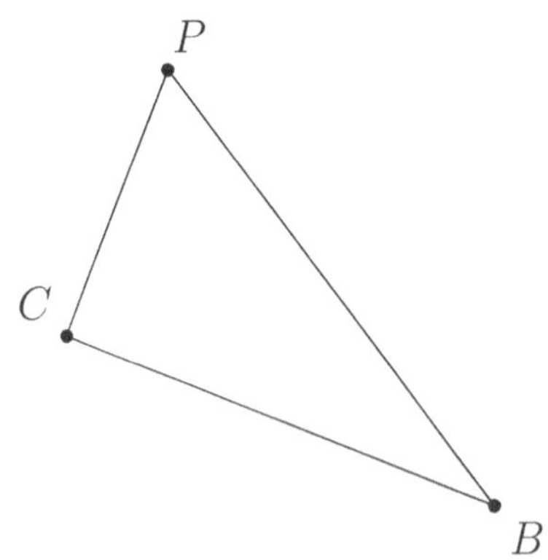

Filling the measures of the angles is tricky in this problem, so let's look at the original diagram again:

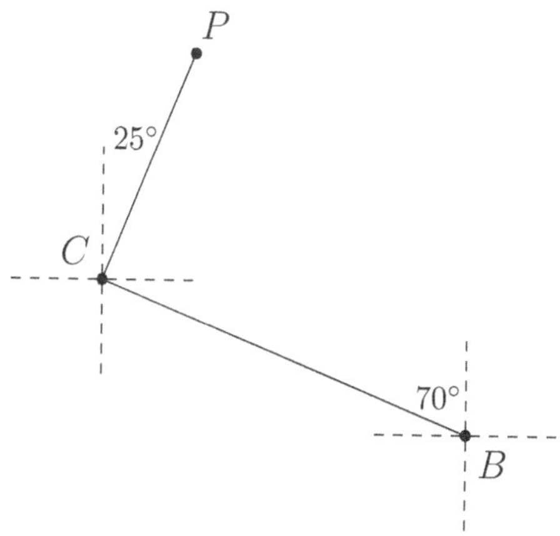

If we extend the dashed line east from Corvallis so that it meets the dashed line running north from Bend, we can create a triangle that shows us that the angle $\angle BCX = 20°$. Also, notice that $\angle PCX = (90° - 25°) = 65°$.

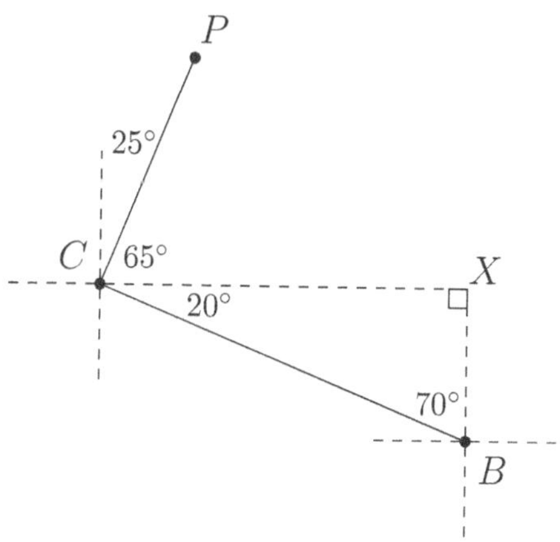

This means that $\angle BCP = 85°$. We know from the problem that $\overline{BC} = 103$ and $\overline{CP} = 72$. We'll need to find the length of $\overline{BP}$ and the measure of $\angle CPB$ to answer the questions.

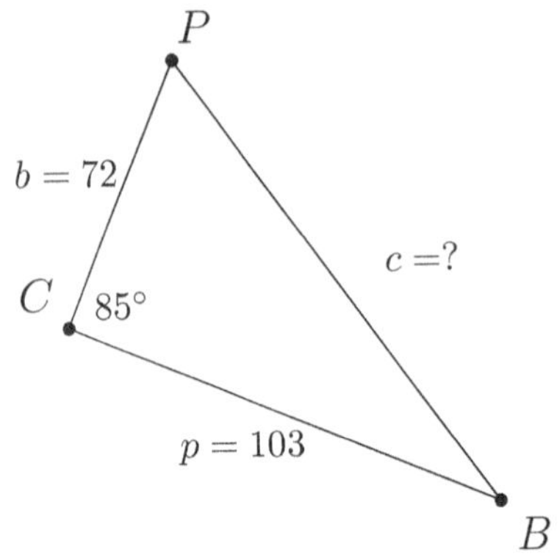

Now we're working with a triangle like the one pictured above, so we can use the Law of Cosines to find the air distance from Portland to Bend:

$$c^2 = b^2 + p^2 - 2bp\cos C$$

$$c^2 = 72^2 + 103^2 - 2 * 72 * 103 * \cos 85^\circ$$

$$c^2 \approx 5184 + 10,609 - 14,832 * (0.087156)$$

$$c^2 \approx 15,793 - 1292.7$$

$$c^2 \approx 14,500.3$$

$$c \approx 120.4 \text{ miles}$$

To find $\angle P$, we'll use the Law of Sines:

$$\frac{\sin 85^\circ}{120.4} = \frac{\sin P}{103}$$

$$103 * \frac{\sin 85^\circ}{120.4} = \sin P$$

$$103 * \frac{0.9962}{120.4} \approx \sin P$$

$$0.85223 \approx \sin P$$

$$58.4^\circ \approx P$$

Now that we know the measure of $\angle P$, we can determine the bearing of Bend from Portland.

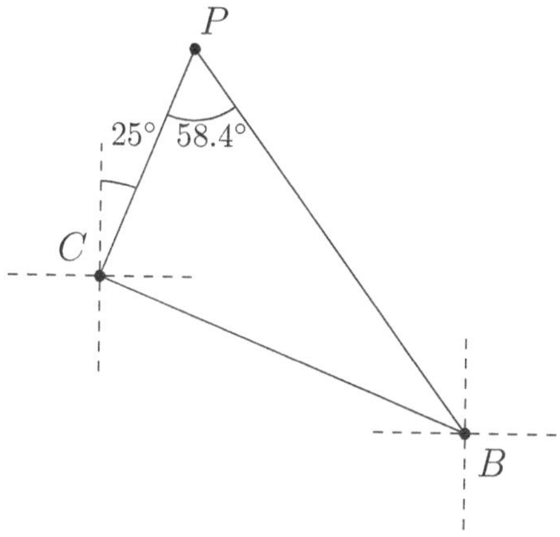

In the picture below notice that $\angle YPC = 25°$. This means that the bearing from Portland to Bend will be east of south by the difference between $\angle P = 58.4°$ and $\angle YPC = 25°$. This makes the bearing of Bend from Portland equal to $S33.4°E$.

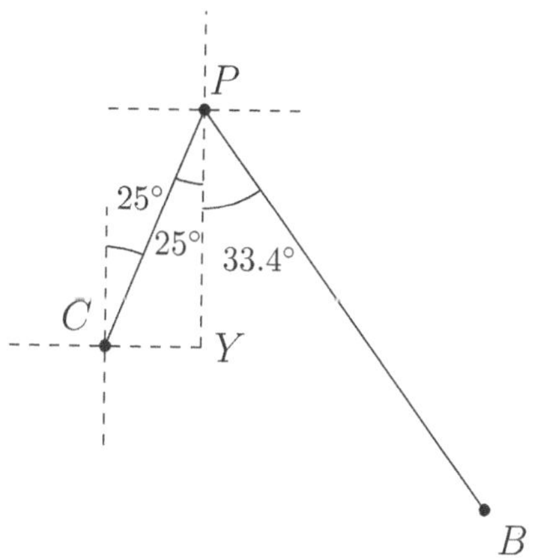

Example 4

A 125 foot tower is located on the side of a mountain that is inclined 32°to the horizontal. A guy wire is to be attached to the top of the tower and anchored at a point 55 feet downhill from the base of the tower. Find the shortest length of wire needed.

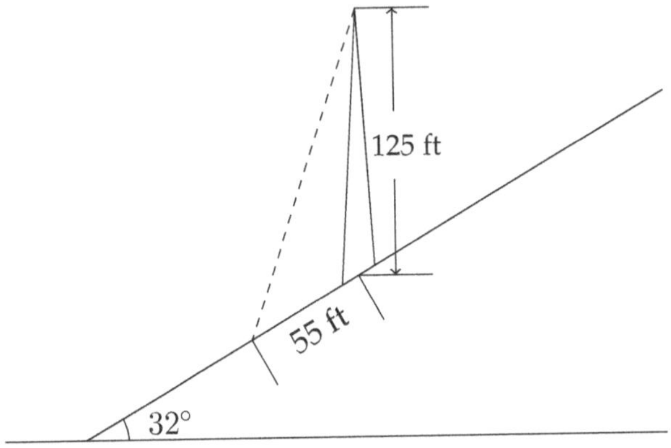

An important aspect in solving this problem is to identify a triangle in the problem which involves the unknown quantity we're being asked to find. If we're looking for the length of the guy wire, we can use a triangle that involves the wire, the distance from the wire to the center of the tower and the height of the tower:

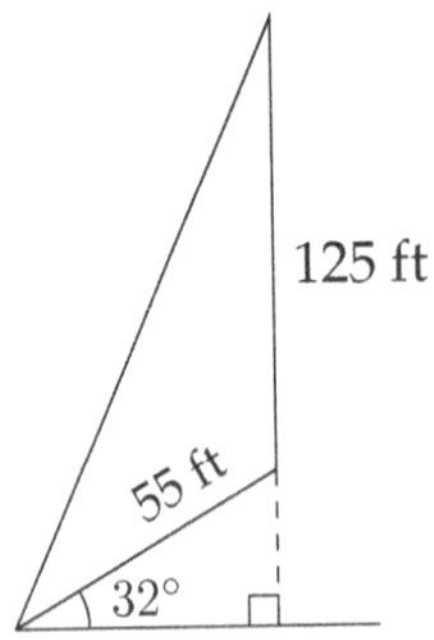

The angle between the horizontal and the hill will stay 32°at any point on the hill. If we drop a perpendicular to the horizontal, we'll be able to find the angle included between the two given sides.

In the little right triangle, we know the 32°angle. That means the other acute angle must be 58°, and the supplementary angle (which is in the triangle we're interested in) will be 122°.

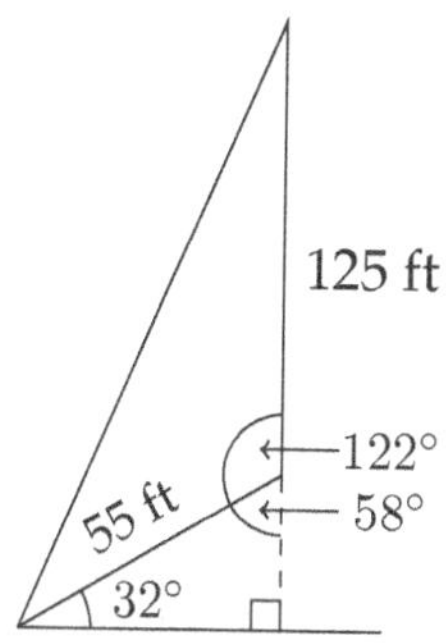

Now we can use the Law of Cosines to find the length of the guy wire:

$$x^2 = 125^2 + 55^2 - 2 * 125 * 55 * \cos 122°$$

$$x^2 = 15,625 + 3025 - 13,750 * \cos 122°$$

$$x^2 \approx 18,650 + 7286.39$$

$$x^2 \approx 25,936.39$$

$$x \approx 161 \text{ feet}$$

Exercises 4.4

1. Two straight roads diverge at an angle of 50°. Two cars leave the intersection at 1 pm, one traveling 60 mph and the other traveling 45 mph. How far apart are the cars (as the crow flies) at 1:30 pm?

2. Two boats leave the same port at the same time. One travels at a speed of 40 mph in the direction $N30°E$ and the other travels at a speed of 28 mph in the direction $S75°E$. How far apart are the two boats after one hour?

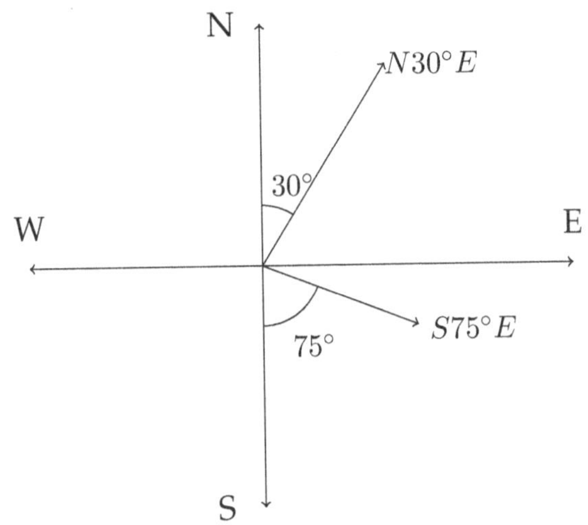

3. The airport in Desert Junction is 350 miles from the airport in Valley Center at a bearing of $N57°E$. A pilot who wants to fly from Valley Center to Desert Junction mistakenly flies due east at 225 mph for 30 minutes before correcting the error. How far is the plane from its destination when the pilot notices the error? What bearing should the plane use in order to arrive at Desert Junction?

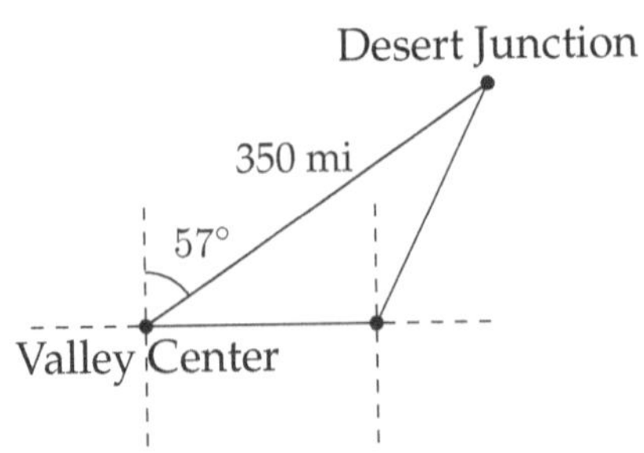

4. An airplane leaves airport A and travels 520 miles to airport B at a bearing of $N35°W$. The plane leaves airport B and travels to airport C 310 miles away at a bearing of $S65°W$ from airport B. Find the distance from airport A to airport C.

5. Two planes take off at the same time from an airport. The first plane flies at 300 mph at a bearing of $S45°E$. The second plane is flying at a bearing of $S5°W$ with a speed of 330 mph. How far apart are they after 3 hours?

6. Two planes leave an airport at the same time. Their speeds are 180 mph and 110 mph, and the angle between their flight paths is 43°. How far apart are they after 2.5 hours?

7. Two ships leave a harbor entrance at the same time. The first travels at a speed of 23 mph and the second travels at 17 mph. If the angle between the courses of the ships is 110°, how far apart are they after one hour?

8. A ship leaves the entrance to a harbor and travels 15 miles with a bearing $S10°W$, then turns and travels 45 miles with a bearing of $N43°W$. How far from the harbor entrance is the ship and what is the bearing of the ship from the harbor?

9. A steep mountain is inclined 77°to the horizontal and rises 3000 feet above the surrounding plain. A cable car is to be installed that will connect the plain to the top of the mountain. The distance from the foot of the mountain to the cable car entry loading area is 1200 feet (see diagram below). Find the shortest necessary length of the cable.

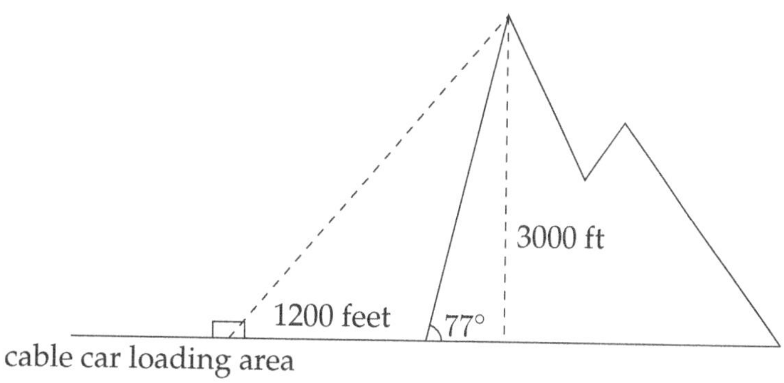

10. A tree on a hillside casts a shadow 208 ft down the hill. If the angle of inclination of the hillside is 25°to the horizontal and the angle of elevation of the sun is 51°, find the height of the tree.

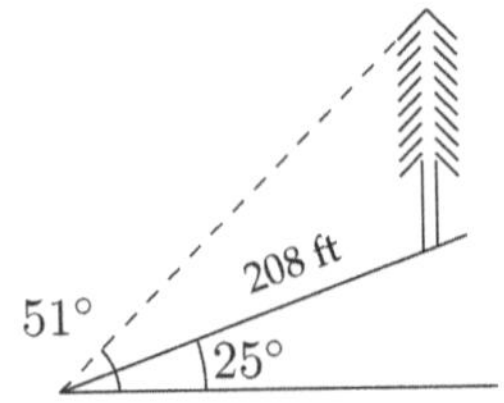

Answer Key

Section 1.1

1. $27.\overline{6}^\circ$
3. $91.8\overline{3}^\circ$
5. 274.3°
7. $17.41\overline{6}^\circ$
9. 183.56°
11. 211.78°
13. $31^\circ\ 25'\ 30''$
15. $6^\circ\ 46'\ 48''$
17. $110^\circ\ 15'$
19. $18^\circ\ 54'$
21. $220^\circ\ 25'\ 48''$
23. $70^\circ\ 12'\ 50.4''$
25. $\frac{\pi}{6}$
27. $\frac{\pi}{4}$
29. $\frac{\pi}{3}$
31. $\frac{\pi}{2}$
33. $\frac{\pi}{12}$
35. $\frac{\pi}{15}$
37. 45°
39. 60°
41. 135°
43. 450°
45. 150°
47. 180°

Section 1.2

1. $\sin\theta = \frac{12}{13}$

   $\cos\theta = \frac{5}{13}$

   $\tan\theta = \frac{12}{5}$

3. $\sin\theta = \frac{8}{17}$

   $\cos\theta = \frac{15}{17}$

   $\tan\theta = \frac{8}{15}$

5. $\sin\theta = \frac{6}{\sqrt{52}} = \frac{3}{\sqrt{13}}$

   $\cos\theta = \frac{4}{\sqrt{52}} = \frac{2}{\sqrt{13}}$

   $\tan\theta = \frac{6}{4} = \frac{3}{2}$

Section 1.2 (cont.)

7. $\sin\theta = \frac{2}{\sqrt{15}}$

$\cos\theta = \sqrt{\frac{11}{15}}$

$\tan\theta = \frac{2}{\sqrt{11}}$

9. $\sin\theta = \sqrt{\frac{3}{7}}$

$\cos\theta = \frac{2}{\sqrt{7}}$

$\tan\theta = \frac{\sqrt{3}}{2}$

11. $\sin\theta = \frac{1}{\sqrt{5}}$

$\cos\theta = \frac{2}{\sqrt{5}}$

13. $\sin\theta = \sqrt{\frac{11}{20}}$

$\tan\theta = \frac{\sqrt{11}}{3}$

15. $\cos\theta = \sqrt{\frac{3}{8}}$

$\tan\theta = \frac{5}{\sqrt{15}}$

17. $\sin\theta = \frac{7\sqrt{31}}{40}$

$\tan\theta = \frac{7\sqrt{31}}{9}$

19. $\sin\theta = \frac{\sqrt{3}}{2}$

$\tan\theta = \sqrt{3}$

21. $\cos\theta = \frac{\sqrt{44}}{7}$

$\tan\theta = \sqrt{\frac{5}{44}}$

Section 1.3

1. Sides: 6, 8, 10
Angles: 36.9°, 53.1°, 90°

3. Sides: 7, 11.03, 13.06
Angles: 32.4°, 57.6°, 90°

5. Sides: 10, 10.72, 14.66
Angles: 43°, 47°, 90°

7. Sides: 16, 25.38, 30
Angles: 32.2°, 57.8°, 90°

9. Sides: 8.47, 33.96, 35
Angles: 14°, 76°, 90°

11. Sides: $\sqrt{11} \approx 3.32$, 7.11, 7.85
Angles: 25°, 65°, 90°

Section 1.4

1. 338.5 feet
3. 33.1°
5. 45.5 feet
7. 343.8 feet, 228.6 feet
9. 6.3 feet, 6.7 feet
11. 32.3 mi., $N59.8^\circ W$
13. 18 mi.
15. 62.8 miles south, 55.8 miles west

Section 1.5

1. 435.5 feet
3. 3278.9
5. 5807.7 feet
7. 149.3 feet
9. 9.9 mi.
11. 106.3 ft.

Section 2.1

1. Quadrant I
3. Quadrant IV
5. Quadrant II
7. Quadrant III

9. Quadrant III
Reference Angle: 15°

11. Quadrant II
Reference Angle: 60°

13. Quadrant II
Reference Angle: 45°

15. Quadrant III
Reference Angle: 80°

17. Quadrant IV
Reference Angle: 45°

19. Quadrant III
Reference Angle: $\frac{\pi}{3}$

21. Quadrant II
Reference Angle: $\frac{\pi}{4}$

23. Quadrant IV
Reference Angle: $\frac{\pi}{6}$

25. $\sin\theta = \frac{3}{5}$

$\tan\theta = -\frac{3}{4}$

27. $\sin\theta = \frac{-3}{\sqrt{13}}$

$\cos\theta = -\frac{2}{\sqrt{13}}$

29. $\cos\theta = -\frac{\sqrt{55}}{8}$

$\tan\theta = -\frac{3}{\sqrt{55}}$

31. $\sin\theta = \frac{5}{\sqrt{26}}$

$\cos\theta = \frac{1}{\sqrt{26}}$

33. $\sin\theta = -\frac{1}{2}$

$\cos\theta = -\frac{\sqrt{3}}{2}$

$\tan\theta = \frac{1}{\sqrt{3}}$

35. $\sin\theta = \frac{1}{\sqrt{2}}$

$\cos\theta = \frac{1}{\sqrt{2}}$

$\tan\theta = 1$

37. $\sin\theta = -\frac{\sqrt{3}}{2}$

$\cos\theta = \frac{1}{2}$

$\tan\theta = \sqrt{3}$

39. $\cos\theta = -\frac{\sqrt{5}}{3}$

$\tan\theta = \frac{2}{\sqrt{5}}$

Section 2.2

1. Amplitude=1, Period=$\frac{\pi}{2}$

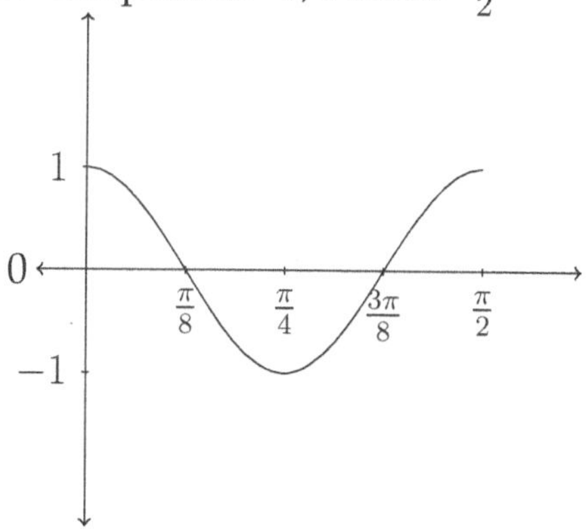

3. Amplitude=3, Period $=\frac{2\pi}{3}$

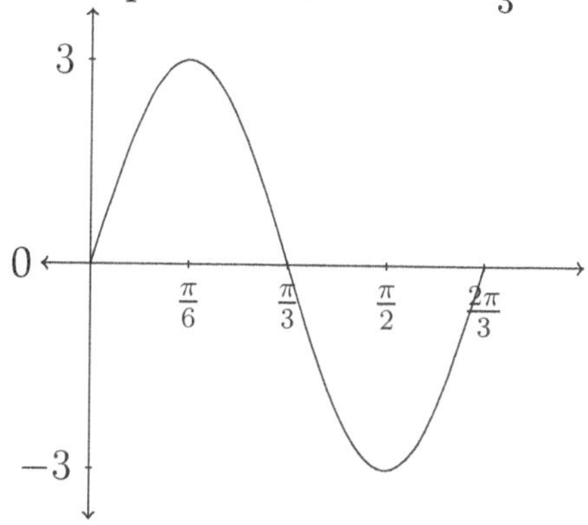

5. Amplitude=4, Period=4π

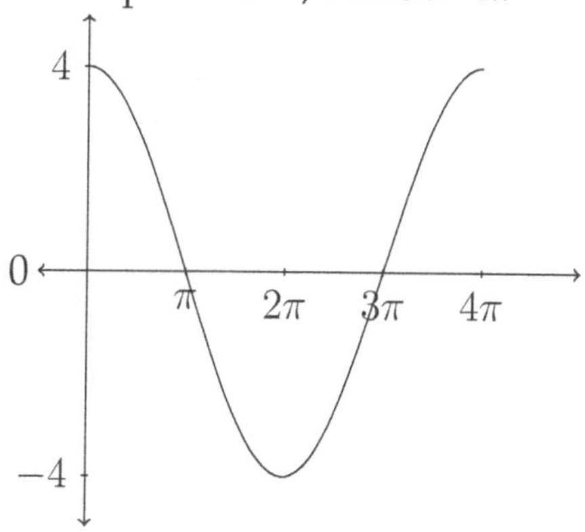

7. Amplitude=$\frac{1}{2}$, Period $=3\pi$

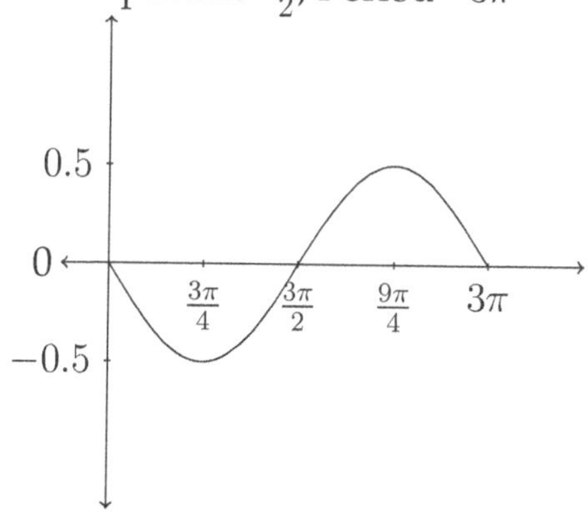

9. Amplitude=4, Period=$\frac{\pi}{3}$

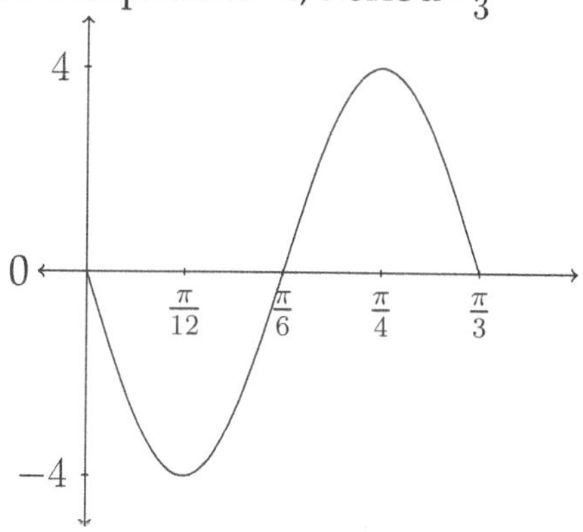

11. Amplitude=2, Period $=\frac{4\pi}{3}$

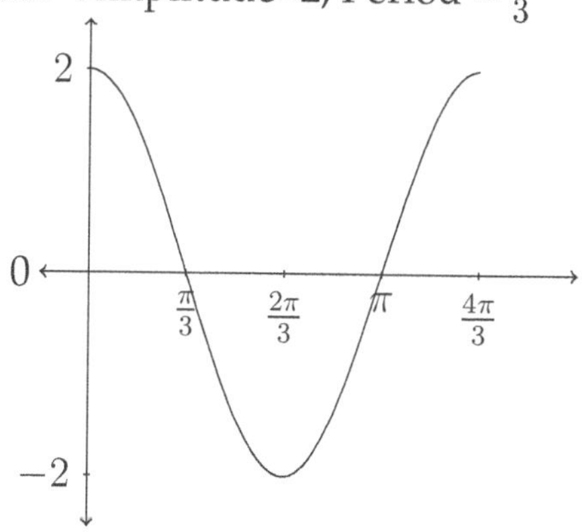

Section 2.2 (cont.)

13. $y = \cos 3x$

15. $y = 4\sin\frac{1}{3}x$

17. $y = -7\cos\frac{4}{3}x$

19. $y = 2\cos\frac{5\pi}{2}x$

Section 2.3

1. Amplitude=1, Period=2π
   Vertical Shift=+1

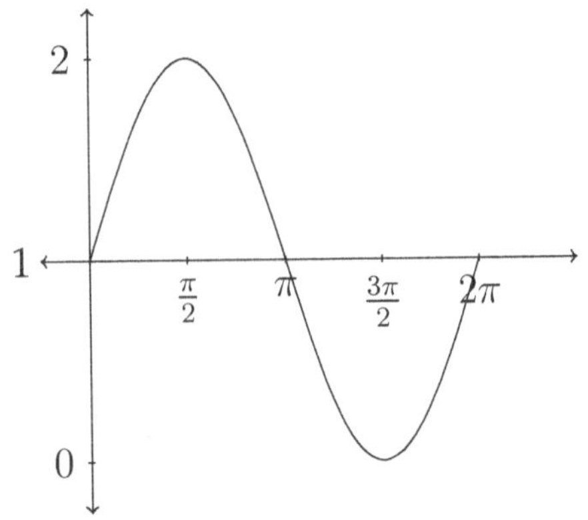

3. Amplitude=2, Period =2π
   Vertical Shift=$-\frac{1}{2}$

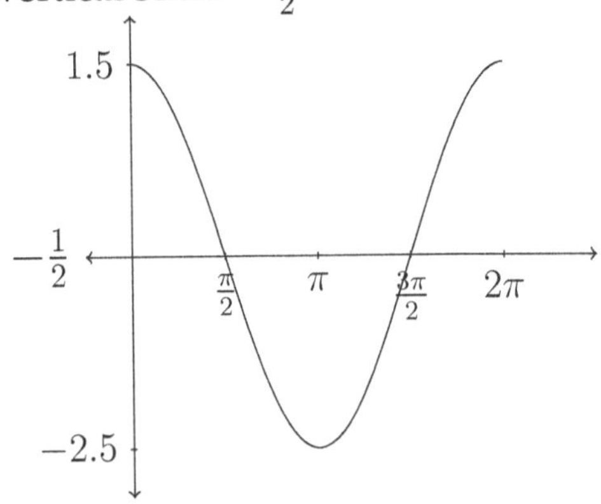

5. Amplitude=1, Period=8π
   Vertical Shift=+1

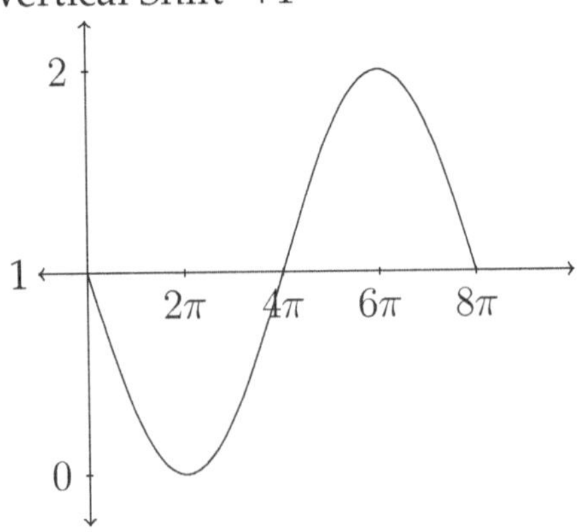

7. Amplitude=$\frac{1}{3}$, Period =2
   Vertical Shift=-4

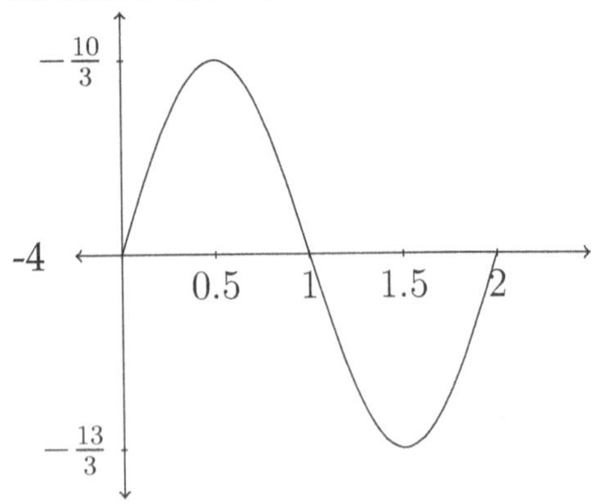

Section 2.3 (cont.)

9. Amplitude=5, Period=4π
   Vertical Shift=+1

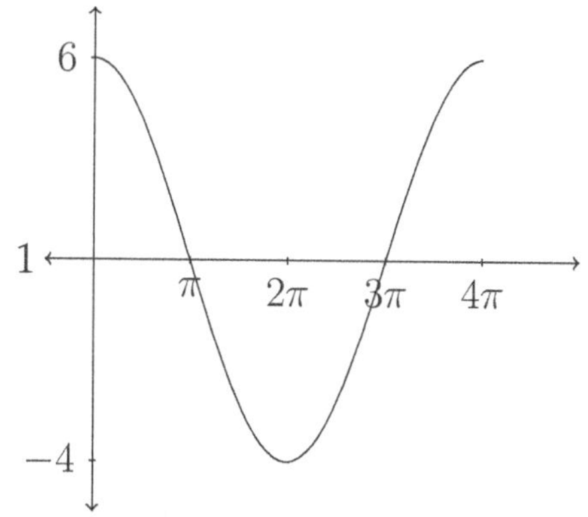

11. Amplitude=3, Period =2π
    Vertical Shift=+2

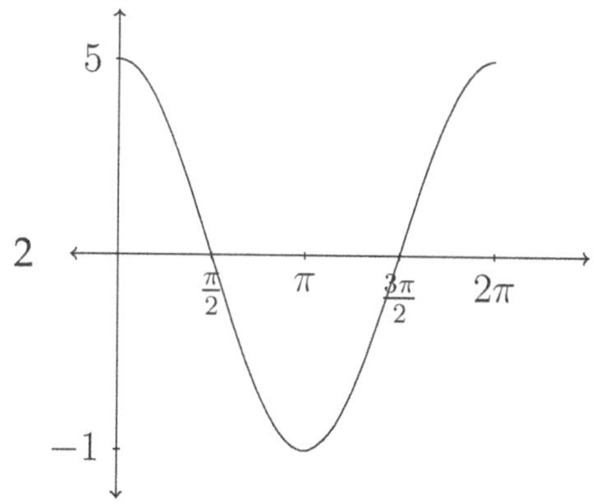

13. Amplitude=4, Period =$\frac{2\pi}{3}$, Vertical Shift=+2

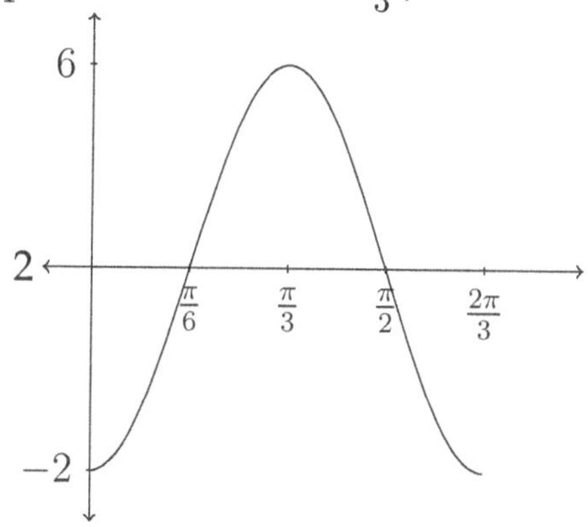

15. $y = 2\cos\frac{5}{2}x + 1$

17. $y = -2\sin\frac{2}{3}x + 7$

19. $y = 2\sin\frac{1}{2}x + 10$

Section 2.4

1. D 2. F 3. E 4. B

5. G 6. C 7. H 8. A

9.

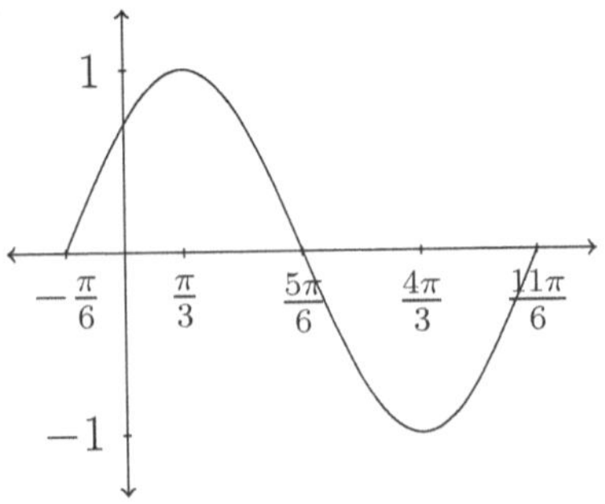

11.

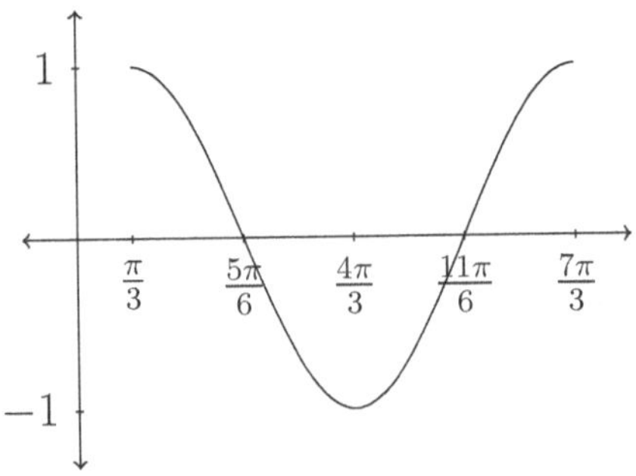

13.

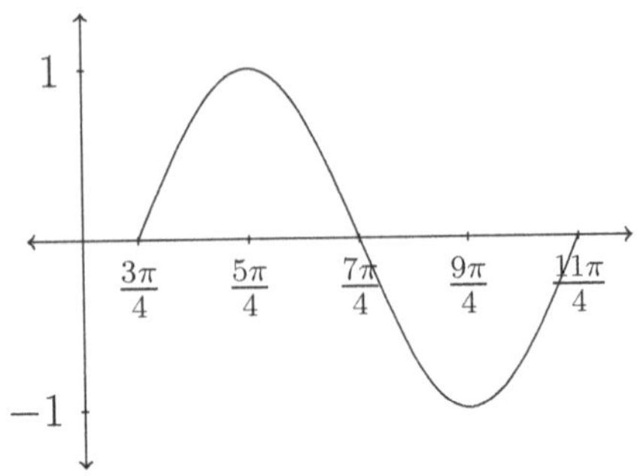

15.

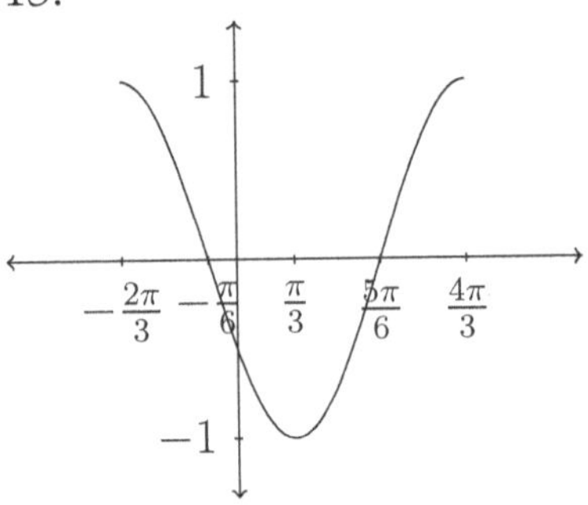

Section 2.5

1. Amplitude=1, Vertical Shift=0
   Period=2π, Phase Shift=$-\frac{\pi}{2}$

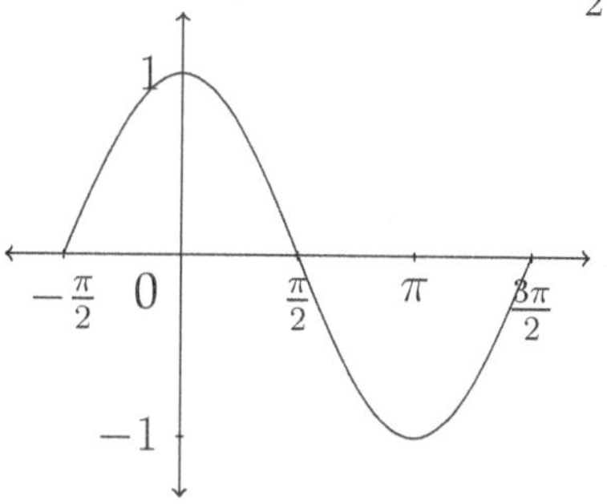

3. Amplitude=3, Vertical Shift=0
   Period =2π, Phase Shift=$\frac{\pi}{2}$

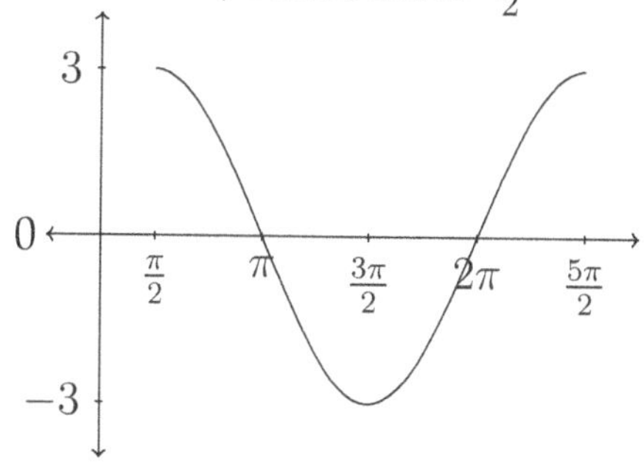

5. Amplitude=1, Vertical Shift=+3
   Period=2π, Phase Shift=$\frac{\pi}{4}$

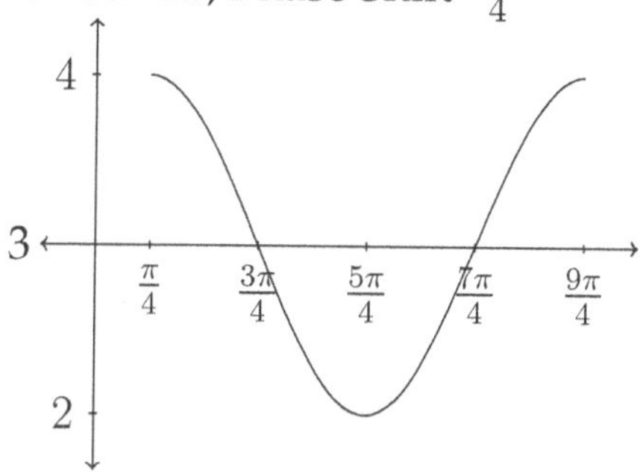

7. Amplitude=1, Vertical Shift=0
   Period =π, Phase Shift=$\frac{\pi}{2}$

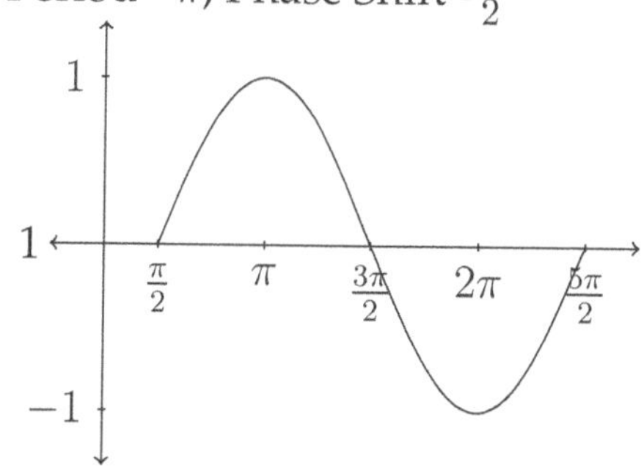

9. Amplitude=2, Vertical Shift=0
   Period=4π, Phase Shift=-2π

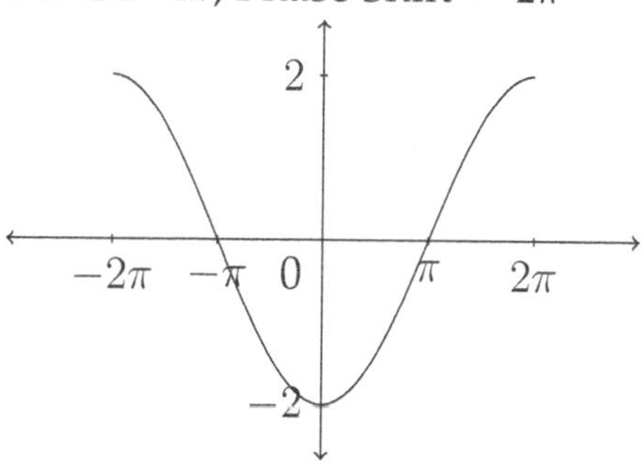

11. Amplitude=$\frac{1}{3}$, Vertical Shift=0
    Period =π, Phase Shift=$-\frac{\pi}{8}$

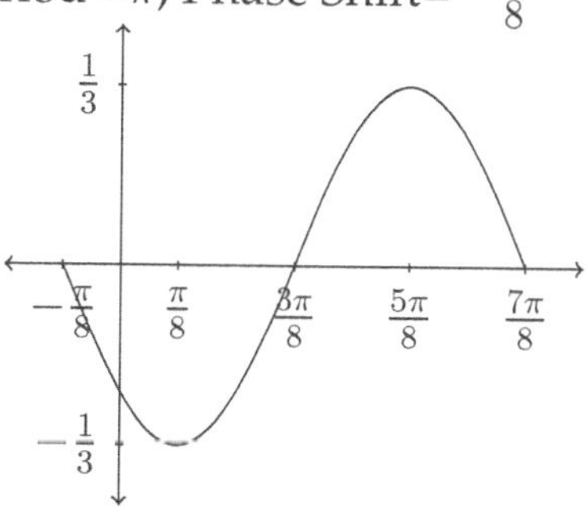

Section 2.5 (cont.)

13. Amplitude=2, Vertical Shift=-1
Period=π, Phase Shift=$\frac{\pi}{6}$

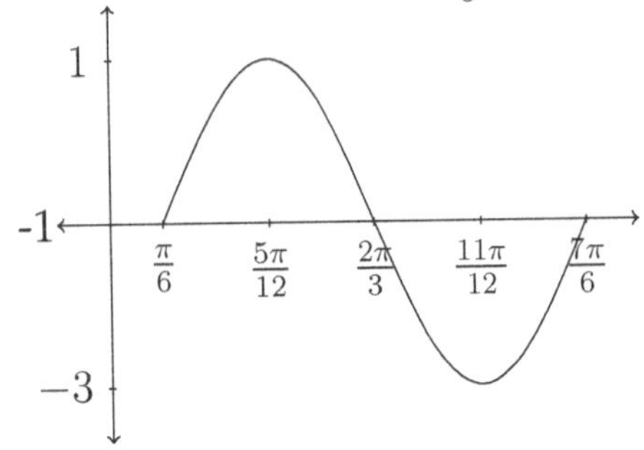

15. Amplitude=3, Vertical Shift=0
Period $=\pi$, Phase Shift=$-\frac{\pi}{6}$

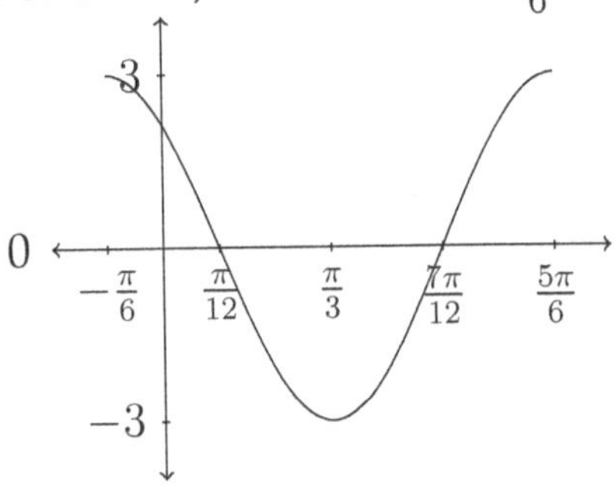

17. Amplitude=1, Vertical Shift=0
Period $=4\pi$, Phase Shift=$-\frac{\pi}{4}$

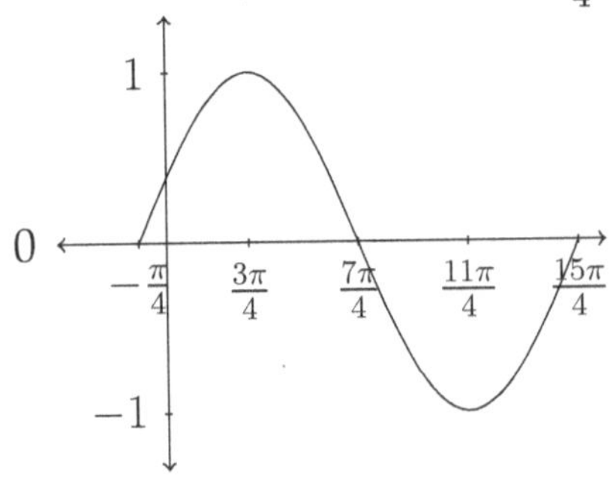

19. $y = 0.5\cos x + 1$

21. $y = -1\sin x - 2$

23. B
24. E
25. H
26. D
27. A
28. G
29. F
30. C

Section 2.5 (cont.)

31. Amplitude=2, Vertical Shift=-1
Period=π, Phase Shift=$-\frac{\pi}{4}$

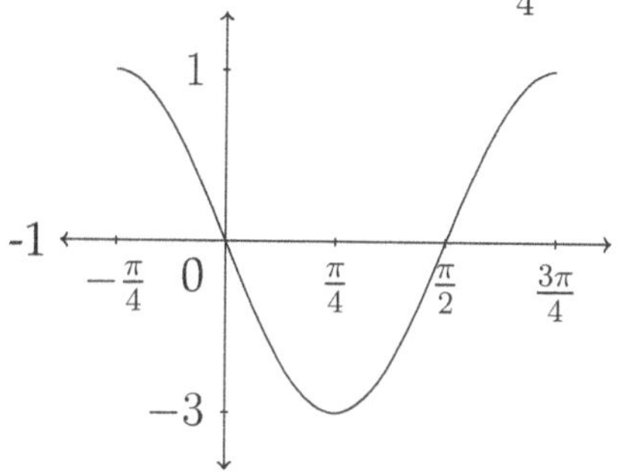

33. Amplitude=1, Vertical Shift=0
Period =π, Phase Shift=$\frac{\pi}{8}$

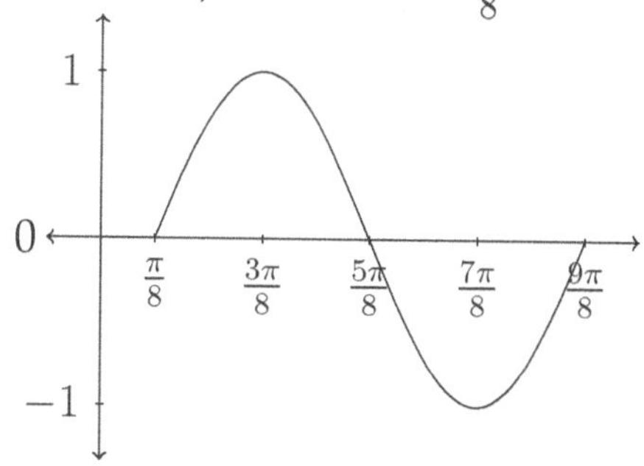

35. Amplitude=3, Vertical Shift=+1
Period=2π, Phase Shift=$-\frac{\pi}{3}$

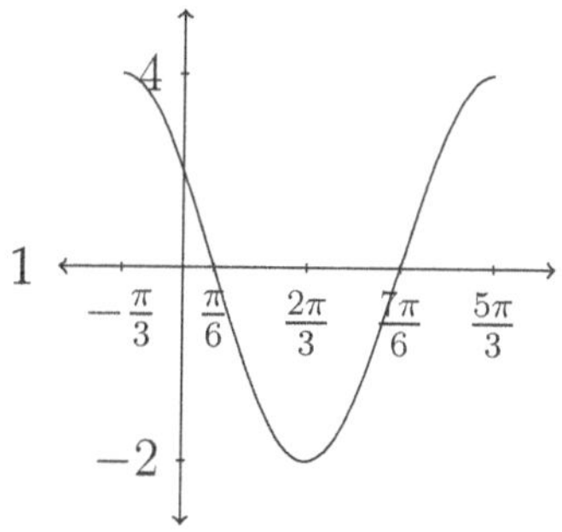

37. Amplitude=$\frac{1}{2}$, Vertical Shift=-2
Period =2π, Phase Shift=$\frac{\pi}{2}$

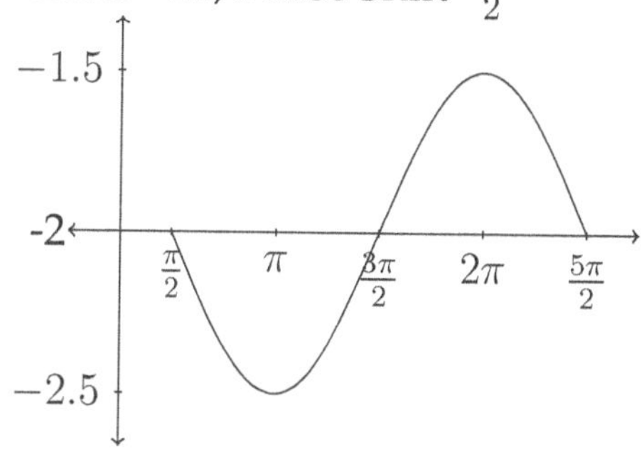

Section 3.3

1. $x \approx 41.4°, 318.6°$
3. no real solution
5. $x \approx 112.0°, 248.0°$
7. $x \approx 12.8°, 167.2°$
9. $x = 0°, 180° \approx 146.3°, 326.3°$
11. $x \approx 70.5°, 289.5°$
13. $x \approx 68.2°, 116.6°, 248.2°, 296.6°$
15. $x \approx 140.0°, 220.0°$

Section 3.4

1. $x \approx 53.6°, 126.4°, 187.9°, 352.1°$
3. $x = 180° \approx 70.5°, 289.5°$
5. $x \approx 49.4°, 130.6°, 190.8°, 349.2°$
7. $x = 180° \approx 41.4°, 318.6°$
9. $x \approx 22.5°, 157.5°$
11. $x = 135°, 315° \approx 71.6°, 251.6°$
13. $x = 0° \approx 138.6°, 221.4°$
15. $x \approx 21.3°, 147.4°, 201.3°, 327.4°$

Section 4.1

1. $\angle A = 25°, \angle B = 30°, \angle C = 125°$
$a \approx 8.45, b = 10, c \approx 16.4$

3. $\angle A = 57°, \angle B = 25°, \angle C = 98°$
$a = 1000, b \approx 503.91, c \approx 1180.76$

5. $\angle A = 102°, \angle B = 28°, \angle C = 50°$
$a = 185, b \approx 88.79, c \approx 144.88$

7. $\angle A = 50°, \angle B = 103°, \angle C = 27°$
$a = 3, b \approx 3.82, c \approx 1.78$

9. $\angle A = 110°, \angle B = 40°, \angle C = 30°$
$a \approx 5.64, b \approx 3.86, c = 3$

11. $\angle A = 23°, \angle B = 110°, \angle C = 47°$
$a \approx 26.71, b \approx 64.24, c = 50$

13. $\angle A = 70°, \angle B = 10°, \angle C = 100°$
$a \approx 10.50, b \approx 1.94, c = 11$

15. $\angle A = 82°, \angle B = 65.4°, \angle C = 32.6°$
$a \approx 39.75, b = 36.5, c \approx 21.63$

17 $\angle A = 42°, \angle B = 61°, \angle C = 77°$
$a = 12, b \approx 15.69, c \approx 17.47$

Section 4.2

1. $\angle A = 50°, \angle B \approx 28.6°, \angle C \approx 101.4°$
$a = 32, b = 20, c \approx 40.95$

3. $\angle A = 43°, \angle B \approx 59.3°, \angle C \approx 77.7°$
$a = 23, b = 29, c \approx 32.95$
$\angle A = 43°, \angle B \approx 120.7°, \angle C \approx 16.3°$
$a = 23, b = 29, c \approx 9.47$

5. no real solution

7. $\angle A \approx 59.8°, \angle B = 24°, \angle C \approx 96.2°$
$a = 17, b = 8, c \approx 19.55$
$\angle A \approx 120.2°, \angle B = 24°, \angle C \approx 35.8°$
$a = 17, b = 8, c \approx 11.51$

9. $\angle A = 108°, \angle B \approx 33.7°, \angle C \approx 38.3°$
$a = 12, b = 7, c \approx 7.82$

11. $\angle A = 42°, \angle B \approx 113.9°, \angle C \approx 24.1°$
$a = 18, b \approx 24.59, c = 11$

Section 4.2 (cont.)

13. no real solution

15. $\angle A = 43°, \angle B \approx 54.5°, \angle C \approx 82.5°$
$a = 31, b = 37, c \approx 45.07$
$\angle A = 43°, \angle B \approx 125.5°, \angle C \approx 11.5°$
$a = 31, b = 37, c \approx 9.06$

17. $\angle A \approx 15.3°, \angle B \approx 32.7°, \angle C = 132°$
$a \approx 7.81, b = 16, c = 22$

19. $\angle A \approx 59.4°, \angle B = 52°, \angle C \approx 68.6°$
$a =\approx 76.46, b = 70, c = 82.7$
$\angle A \approx 16.6°, \angle B = 52°, \angle C \approx 111.4°$
$a \approx 25.38, b = 70, c = 82.7$

Section 4.3

1. $\angle A \approx 52.5°, \angle B = 30°, \angle C \approx 97.5°$
$a = 24, b \approx 15.13, c = 30$

3. $\angle A \approx 47.8°, \angle B \approx 21.3°, \angle C \approx 110.9°$
$a = 122, b = 60, c = 154$

5. $\angle A = 52°, \angle B \approx 80.3°, \angle C \approx 47.7°$
$a \approx 28.78, b = 36, c = 27$

7. $\angle A \approx 17.8°, \angle B = 135°, \angle C \approx 27.2°$
$a = 12, b \approx 27.81, c = 18$

9. $\angle A \approx 50.3°, \angle B \approx 72.3°, \angle C \approx 57.4°$
$a = 21, b = 26, c = 23$

11. $\angle A \approx 38.7°, \angle B \approx 53.1°, \angle C \approx 88.2°$
$a = 25, b = 32, c = 40$

13. $\angle A = 77.4°, \angle B \approx 55.9°, \angle C \approx 46.7°$
$a \approx 523.15, b = 444, c = 390$

15. $\angle A \approx 57.4°, \angle B \approx 46.1°, \angle C \approx 76.5°$
$a = 112.7, b = 96.5, c = 130.2$

Section 4.4

1. 23.2 miles

3. 262.9 miles, $N43.5°E$

5. 802.9 miles

7. 32.9 miles

9. 3547.1 feet

INDEX

F

G

H

I

L

M

N

O

P

Q

R

9 788197 559426

Printed at Repro India Ltd.